Cybersecurity in Knowledge Management

In an era where digital transformation is vital across industries, protecting knowledge and information assets has become critical. *Cybersecurity in Knowledge Management: Cyberthreats and Solutions* explores the intersection of knowledge management and cybersecurity, offering an in-depth examination of the strategies, technologies, and frameworks necessary to safeguard organizational knowledge systems. As cyber threats grow more sophisticated, particularly within sectors such as digital marketing, supply chains, and higher education, this book examines methods for enhancing cybersecurity while maintaining the agility needed to foster innovation. By incorporating perspectives from artificial intelligence, machine learning, and human factors, this work provides a holistic approach to securing knowledge in today's interconnected landscape.

This book includes an analysis of AI and machine learning applications for cybersecurity, a comparative review of malware classification techniques, and real-world case studies illustrating cybersecurity breaches and insider threats affecting knowledge ecosystems. This book addresses unique challenges within the African digital space, explores social engineering tactics, and emphasizes the role of organizational culture in maintaining knowledge security. Key topics include cybersecurity requirements in digital marketing, the post-COVID impact on knowledge transfer in higher education, and the importance of regulatory compliance and cross-industry collaboration.

With its multidisciplinary perspective, *Cybersecurity in Knowledge Management: Cyberthreats and Solutions* is ideal for professionals, researchers, and policymakers. This comprehensive guide equips readers with the insights needed to build resilient cybersecurity programs that protect essential knowledge assets, enabling organizations to meet today's cybersecurity demands while maintaining a sustainable competitive advantage in an evolving digital environment.

Cybersecurity in Knowledge Management

Cyberthreats and Solutions

Edited by

Narasimha Rao Vajjhala

and

Kenneth David Strang

CRC Press
Taylor & Francis Group
Boca Raton London New York

CRC Press is an imprint of the
Taylor & Francis Group, an **informa** business

Designed cover image: Shutterstock ID: 2353592457

First edition published 2025
by CRC Press
2385 NW Executive Center Drive, Suite 320, Boca Raton FL 33431

and by CRC Press
4 Park Square, Milton Park, Abingdon, Oxon, OX14 4RN

CRC Press is an imprint of Taylor & Francis Group, LLC

ISBN: 978-1-032-80231-2 (hbk)
ISBN: 978-1-032-80677-8 (pbk)
ISBN: 978-1-003-49809-4 (ebk)

DOI: 10.1201/9781003498094

Typeset in Times
by SPi Technologies India Pvt Ltd (Straive)

Contents

Editors

Narasimha Rao Vajjhala

American University in Bulgaria, Bulgaria

Narasimha Rao Vajjhala is an academic and researcher currently serving as Professor and Chair of the Computer Science Department at the American University in Bulgaria (AUBG). He is also the Principal Investigator for the Democracy Project (DemPro) at the Center for Information, Democracy, and Citizenship (CIDC) at AUBG. Dr. Vajjhala previously held leadership roles as Dean of the Faculty of Engineering and Architecture at the University of New York Tirana (UNYT) in Albania and Chair of the Computer Science and Software Engineering programs at the American University of Nigeria (AUN). With over 23 years of experience in academia, he has taught programming, database management, and related courses across Europe and Africa at both undergraduate and graduate levels. An active contributor to the global research community, Dr. Vajjhala is a Senior Member of ACM and IEEE, Editor-in-Chief of the *International Journal of Risk and Contingency Management (IJRCM)*, and a member of the Project Management Institute (PMI). Beyond academia, he has worked as a technology consultant for European firms and has been actively involved in EU-funded projects. Dr. Vajjhala holds a Doctorate in Information Systems and Technology from the United States, a Master of Science in Computer Science and Applications from India, and a Master of Business Administration (MBA) specializing in Information Systems from Switzerland. His expertise spans computer science, project management, and information systems, making significant contributions to both research and industry.

Kenneth David Strang, PhD, MBA, BS, AS, CNA, CSCS, CRP, CPP, FLMI, PMP

W3-Research, USA and RMIT University, Australia, University of the Cumberlands, Williamsburg, USA

Dr. Strang has over 330 publications across multiple disciplines (http://kennethstrang.com) and at the time of writing he has consistently ranked in the 95–97 percentiles of worldwide scholars at Research Gate. Dr. Strang has been working since 1983. He has been employed by W3-Research, a business analytics company, since 2010. He is a retired full professor who occasionally lectures at reputable universities and associations. He has a professional Doctorate in project management operations research, an MBA in strategic management, a BS in marketing, and an AS in computer science all with summa cum laude high-grade distinctions.

He has six professional certifications/licenses: Certified Netware Administer (security), a Fellow Life Management Institute (insurance actuary), a Certified Research Professional (scientific research), a Project Management Professional, a Certified Supply Chain Specialist/Procurement Professional, and a World Medical Association research ethics certification. Dr. Strang conducts research in business, healthcare, logistics/supply chain, manufacturing, marketing, mining, new product development, leadership, organizational behavior, industrial psychology, tourism marketing, human resource management, learning styles, online learning, nuclear-hydro-solar-wind clean energy production, and procurement project management. He has edited a best-selling research design-methods textbook for Palgrave-Springer and he won over $8M USD in grant proposals to support his clients. He also won two Emerald Literati Awards. He has managed over 115 studies and taught over 1000 multicultural students with an anonymous course opinion survey satisfaction median of 5/5 and a mean of 4.5/5. Dr. Strang also volunteers for many charities and not-for-profit social advocate associations.

Contributing Authors

Adeyemi Abel Ajibesin is a distinguished academic scholar with over two decades of extensive experience in research, teaching, and leadership, particularly in Electrical Engineering and Computer Science. Currently serving as a visiting professor in the Department of IT at Cape Peninsula University of Technology in South Africa, he previously held positions such as Chair/HoD of Software Engineering at the School of IT & Computing (SITC) at the American University of Nigeria, founding Interim Dean of the School of Engineering, Interim Dean of the School of IT & Computing, and Director of the African Centre for ICT Innovation and Training. He has also served in various capacities at Simon Fraser University in Canada, the University of Cape Town in South Africa, and the Pan African University Institute for Basic Sciences, Technology Innovation in Kenya, among other institutions. His areas of research are data science, broadband networks, frontier analysis, and AI. His prolific research portfolio is reflected in numerous articles published in reputable peer-reviewed conference proceedings, journals, and books, alongside patents for engineering products and copyrighted computing software, showcasing his innovative contributions to his field. Dr. Ajibesin's leadership and academic excellence have been recognized through numerous awards and grants, including esteemed accolades such as the Google Research (explorerCSR), AIMS, CSIR, and IEEE.

Junnussov Aruan is a researcher at the KBTU Business School of the Kazakh-British Technical University of Kazakhstan, working primarily in management and marketing. Mr. Aruan has been continuing his research, mostly marketing, CSR, and Sustainability, focusing on Knowledge and Innovation Management areas.

Arti Arya holds a B.Sc. (Mathematics Hons) and an M.Sc. (Mathematics) from Delhi University, completed in 1994 and 1996, respectively. She earned her Ph.D. in Computer Science Engineering from the Faculty of Technology and Engineering, Maharishi Dayanand University, Rohtak, Haryana, in 2009. Currently, she serves as a Professor in the Department of Computer Science Engineering at PES University, Bengaluru, India, bringing 24 years of academic experience, including 16 years dedicated to research. Her expertise spans spatial data mining, knowledge-based systems, text mining, computational linguistics, machine learning, artificial intelligence, applied numerical methods, Generative AI, and biostatistics. Dr. Arya is a life member of the Computer Society of India (CSI) and a member of IEEE, and she serves on the review boards of several prestigious international journals.

Mukashev Erzhan Bulatovich serves as the Dean and Associate Professor at the School of Business, Kazakh British Technical University, Kazakhstan. He holds a Ph.D. in Economics from the University of Maryland, USA, and brings over two decades of dedicated research experience. Dr. Erzhan has an extensive publication record, including research papers, book chapters, and presentations at international conferences within his areas of expertise. His global perspective and rich experience in multicultural and multi-ethnic environments enhance his effectiveness in fostering international collaborations and managing organizational administration. His research interests include Comparative Institutional Economics, Macroeconomics, Project Management, Entrepreneurship, and Small Business Management.

Eriona Çela is a leading expert in Albania in the fields of cybersecurity, knowledge management, artificial intelligence, and deep learning, recognized for her significant impact on education and public administration within Albania and beyond. As a distinguished academic, she holds prominent roles across multiple institutions, including the Academy of European Integration and Negotiations (AIEN), the Albanian School of Public Administration (ASPA), and the Business Academy College. Her contributions are widely acknowledged in areas ranging from advanced technologies to European Union policies and enlargement strategies, where she is known for her expertise in fostering digital and technological advancement within educational frameworks. Dr. Çela's role as a Jean Monnet Chair at the Institute of European Studies, University of Tirana, underscores her leadership in European integration, having spearheaded major initiatives such as the eighth edition of the European Integration School in 2023. Her research is frequently published in esteemed journals, including the European Journal of Arts, Humanities and Social Sciences, the European Journal of Contemporary Education, and the International Journal of Balkan Studies. Additionally, as an Associate Editor for the International Journal of Risk and Contingency Management (IJRCM), an IGI Global Journal, she further contributes to global scholarship in her fields of expertise. Dr. Çela has advanced her research on European integration and technology applications through collaborations with institutions such as the OSCE Academy in Bishkek, Vytautas Magnus University, and the Università degli Studi di Bari Aldo Moro. Her dedication to educational leadership is exemplified by her work with Shkolla e Drejtorëve (CSL) and the AADF, where she focuses on nurturing future leaders. A frequent keynote speaker at academic and policy-making events, Dr. Çela continues to shape the strategic direction of educational and technological standards in Albania, setting a benchmark for integrating European practices in the region.

Rahul Dasari is a dedicated researcher with a strong foundation in technology, computer science, and cybersecurity. Mr. Dasari has demonstrated a commitment to both academics and extracurricular activities. His research contributions showcase his expertise in cybersecurity. Mr. Dasari has also completed Stanford Online's Statistical Learning with Python and actively participates in the National English Honor Society, National Honor Society, and competitive Esports. With experience in coding, PC building, and robotics, he combines technical skills with a passion for innovation and education, evidenced by his role as an educator at Code Ninjas.

Basil Osayin Daudu holds a Ph.D. in Social and Political Philosophy with a specialization in African Studies, focusing on societal, political, and cultural dimensions. He is a licensed instructor and a certified Microsoft Innovative Educator, contributing regularly to academic journals and book chapters. Currently, he serves as a lecturer and researcher in the Department of Philosophy at Kogi State University, Anyigba, Nigeria.

Dwarampudi Aiswarya has served as an Assistant Professor in the Department of Computer Science and Engineering at Pragati Engineering College (Autonomous), Surampalem, bringing nearly two years of teaching experience. She has published over 25 research papers in esteemed national and international journals. Her research interests focus on cybersecurity and machine learning. Additionally, she has authored a book on cybersecurity and holds a patent in this field, reflecting her contributions to advancing knowledge and innovation in cybersecurity.

Philip Eappen is an Assistant Professor of Healthcare Management at Cape Breton University. He is also the Director of Clinical Services and Transition to Community at the Breton Ability Center (LOA). With over a decade of experience in hospital and healthcare management operations, Dr. Eappen has a strong academic background. Before joining Cape Breton University, he taught healthcare management at the University of Toronto, the Southern Alberta Institute of Technology, and Fanshawe College. Dr. Eappen also worked as a Director of Health Services and Chief Administrator of Healthcare Operations at the American University of Nigeria. Additionally, he has served as a faculty member at the American University and other universities and colleges. Dr. Eappen holds a doctorate in healthcare administration from Central Michigan University in the United States, a Master of Business Administration in Healthcare Management, a Bachelor of Nursing, and a Post-Graduate Certificate in International Health. Dr. Eappen is a board director on various boards, including the American College of Healthcare Executives, Myeloma Canada, and Aplastic Anemia and Myelodysplasia Canada. He is also a long-serving member of the American College of Healthcare Executives and the Canadian College of Health Leaders. His research interests are focused on healthcare innovations, particularly healthcare informatics. In recent years, Dr. Eappen has published several articles, book chapters, and books on the subject.

Mathias Mbu Fonkam is an accomplished Associate Teaching Professor at Penn State University with over 20 years of experience in computer science education and as a practicing software engineer. He holds a Ph.D. in Computer Science, an M.Sc. in Systems Engineering and a BSc in Computer Engineering from Cardiff University, UK. His professional journey includes serving as the Dean at the American University of Nigeria (AUN), where he spearheaded numerous initiatives, such as developing the university's Open-Source ERP System and establishing various postgraduate programs. Dr. Fonkam's research interests span AI, data science, blockchain, open-source software and systems dynamics, with over 30 publications in reputable conferences and journals. He has also contributed significantly to curriculum development and accreditation processes in Nigerian universities. Fluent in English, Portuguese, and French, Dr. Fonkam is dedicated to advancing computing education and professional development, leveraging his extensive expertise in teaching, mentoring, and research.

R. Sankar Ganesh, is an accomplished academic currently serving as an Associate Professor at Vel Tech Rangarajan Dr. Sagunthala R&D Institute of Science and Technology, Chennai. With expertise in marketing and human resource management, Dr. Ganesh brings over 11 years of teaching experience alongside four years in industry. He holds two design patents registered with IPR India and has contributed extensively to the academic community as a resource person at various conferences and faculty development programs (FDPs). Additionally, Dr. Ganesh has presented papers at numerous forums, participated in both national and international seminars, and published research in Scopus, Web of Science, and UGC Care-listed journals. His commitment to mentorship is reflected in his guidance of Ph.D. scholars and postgraduate students in their research pursuits.

Ganesh Khekare holds a Doctor of Computer Science and Engineering from Bhagwant University, India in 2021. He is a postdoc fellow from Lincoln University, Malaysia. He also received his B.E. and M.E. (CSE) from Nagpur University, India in 2010 and 2013, respectively. He is currently working as an associate professor at the Computer Science & Engineering Department at Vellore Institute of Technology, Vellore, India. His research areas include Artificial Intelligence and Machine Learning, Data Science, Networks, Internet of Things, etc. He has published over 70 papers in international journals and conferences. He has done 5 Patents and 10 Copyrights. He is an active member of various professional societies like ACM, ISTE, IEEE Senior Member, IEI, etc.

Turarbekova Kuralay is a business disciplines researcher at the Business School of the Kazakh-British Technical University of Kazakhstan. Ms. Kuralay is voracious and extensively researching knowledge management, artificial intelligence, and Sustainability, mainly in the service and education sectors, and is currently working on two significant projects considered spheres of business development.

M. V. Sujan Kumar is an Integrated MTech Computer Science research scholar at the Vellore Institute of Technology. He has developed projects such as the Course Records Portal using Python and MySQL. Sujan gained experience in data analysis, modeling, Java, and web development during his internships as a Data Science Intern and Java Software Engineer Intern. His research interests focus on enhancing fake news detection integrity, leveraging generative AI for brain tumor detection, and chronic kidney tumor detection using ML-DL. An active member of Dijkstra-ML Open Source, he holds certifications from CISCO and Coursera. He is skilled in Machine Learning, Deep Learning, Python, Java, SQL, Jenkins, CI/CD, and Front-End Development.

Swati Maste is a dedicated researcher pursuing her Master of Science in Artificial Intelligence (MSAI) at the University of Texas at Austin. She holds a Bachelor of Technology (B.Tech.) in Computer Science, specializing in Data Science and Machine Intelligence, from PES University, completed in 2023. Swati has contributed to the field with two published research papers: "Multi-step Online Hate Speech Detection and Classification Using Sentiment and Sarcasm Features" and "mURLi: A Tool for Detection of Malicious URLs and Injection Attacks." Currently, she is a software developer at SAP Labs India, where she applies AI-driven solutions to address complex business challenges. Her research interests focus on advancing Natural Language Processing (NLP), Machine Learning (ML), Deep Neural Networks, and Image Processing.

Goddy Uwa Osimen is a PhD holder in Strategic Studies and currently serves as a Senior Lecturer and Research Fellow in the Department of Political Science and International Relations at Covenant University, Ota, Nigeria. He has an extensive portfolio of scholarly publications, with research interests centered on International Politics, Warfare, and Strategic Studies.

Ritu Pareek is an Assistant Professor of English at Birla Institute of Technology, Mesra, Jaipur Campus, with over 15 years of academic experience. She earned her PhD from Malaviya National Institute of Technology, Jaipur, in 2014. A certified Cambridge English instructor specializing in Communication Skills, Dr. Pareek has contributed significantly to her field through published articles on Contemporary British Fiction and English Language Teaching in esteemed journals. She has also presented research at numerous international conferences, including the University of London, UK, and at several national conferences across Indian universities. Additionally, Dr. Pareek has translated various NPTEL courses and collaborates with Svitlo School, a UK-based charitable organization supporting Ukrainian children.

Rajasekhara Mouly Potluri is a distinguished Professor of Management and Marketing with over thirty-three years of industry and academic experience across ten countries. Currently a Professor of Marketing at the Business School of Kazakh British Technical University in Almaty, Kazakhstan, Dr. Potluri holds a Ph.D. and M.Phil. in Management and Marketing from Shivaji University, along with an MBA in Marketing and an M.Com. in Banking from Andhra University, India. He has authored approximately 130 research articles, books, book chapters, and case studies, with publications in prestigious peer-reviewed journals indexed by Scopus, ABDC, SSCI, IEEE, and KCI, as well as presentations at international conferences. His work has earned him over twenty awards for Best Research Paper and Academic Service Excellence. Prof. Potluri's extensive teaching and research experience spans diverse multicultural and multi-ethnic settings, including India, Ethiopia, Kazakhstan, South Korea, Nigeria, and the United Arab Emirates, where he has received a Golden Visa. His research interests include Marketing, Islamic Marketing, CSR and Sustainability, HRM & OB, and Entrepreneurship. He is an active member of several professional organizations, including the Eurasian Business & Economics Society, the International Islamic Marketing Association, the Social Responsibility Research Network, the International Society for Development & Sustainability, the Emerald Literati Network, and the Consumer Guidance Society of India.

Lohith Sekhar Potluri is a dedicated Data Scientist at the School of Medicine, University of California San Diego, USA. With a master's degree in data science, Mr. Potluri is a prolific researcher in his field, with multiple publications, including research articles, book chapters, and conference presentations focused on data science, machine learning, artificial intelligence, and statistics. He possesses strong expertise in R Language, R Studio, SQL, and SaaS. Recognized for his contributions, he has received four Best Paper and Academic Excellence awards at international conferences and is currently engaged in three research projects.

Pallavi Prabhu is a researcher with a B.Tech. in Computer Science and Engineering, specializing in Machine Intelligence and Data Science, from PES University, Bangalore. Her research is centered on the intersection of Natural Language Processing and Machine Learning, where she explores advanced methodologies to tackle pressing challenges in the field. Pallavi has published impactful work on Hate Speech Detection, leveraging cutting-edge technologies to address complex, real-world issues.

Lukman Raimi is an Assistant Professor of Entrepreneurship and Programme Leader at Universiti Brunei Darussalam. He has previously served at the American University of Nigeria in various leadership roles, including Chair of Entrepreneurship and Deputy Director of the Center for Entrepreneurship. His editorial expertise includes serving on the boards of prestigious journals such as the Journal of Developmental Entrepreneurship (USA), Emerging Markets Case Studies (UK), Editorial Board, Indonesian Journal of Sustainability Accounting & Management (Indonesia), and the Journal of Green Economy and Low-Carbon Development (Romania). Raimi is also an Associate Editor of the Journal of Business and Economic Analysis at Universiti Brunei Darussalam. His experience spans peer reviewing for notable journals like the Journal of Global Entrepreneurship Research and the Journal of Business Ethics. His work emphasizes entrepreneurship, ethical business practices, sustainability, and corporate social responsibility, and he has cultivated a vast network of international scholars, enriching his contributions to academic and professional discourse on entrepreneurship and innovation across continents.

Anurup Sankriti is a Software Engineering M. Tech. Integrated research scholar at Vellore Institute of Technology with a strong interest in Machine Learning and Neural Networks. Proficient in Python, Java, C++, and other languages, he is enthusiastic about coding and implementing innovative ideas. In addition to his academic pursuits, Anurup values collaboration and effective communication. He recently interned at Cognerium, where he developed predictive models, machine learning algorithms, and deep learning models. An ML Open-Source Member at Dijkstra, he also contributes articles on ML techniques. With expertise in programming, databases, ML/NLP packages, and software testing, Anurup combines technical skills with a sociable approach to his projects.

Rakesh Kumar Saxena holds an M.Tech. in Engineering Systems from DEI Agra (1999) and a Ph.D. in Computer Architecture and Logic Design from RGPV Bhopal (2016). Dr. Saxena is a Professor of Computer Science at Poornima University, Jaipur. He is a Fellow of the Institution of Engineers India, IACSIT, and ISTE Delhi, and was awarded Chartered Engineer status by IE (I) in 2017. Dr. Saxena's research primarily focuses on soft computing, including Artificial Intelligence, Machine Learning, and Deep Learning. He has successfully led projects funded by AICTE, MSME, and the World Bank and has published 37 research papers in IEEE Xplore, AIP, and other Scopus/SCI-indexed journals.

S. Senthilkumar is an Associate Professor in the Department of Management at Skyline University, Nigeria. He holds an MBA, M.Com, M.S., and Ph.D., and brings 22 years of extensive experience in teaching and academia. Dr. Kumar has published 71 papers in national and international journals and authored four books on Research Methodology, Corporate Finance, Principles of Accounting, and Business Policy & Strategic Management. He has presented 70 papers at various national and international seminars and conferences, in addition to serving as a resource person and guest lecturer in approximately 25 programs. Dr. Kumar has guided numerous Master's and Ph.D. students and is an authorized Ph.D. and M.Phil. supervisor for both Indian and foreign universities. He has also acted as a foreign examiner, evaluating around

75 Ph.D. theses for universities worldwide. An active reviewer and editorial board member, Dr. Kumar contributes his expertise to journals focused on Commerce, Management, and Entrepreneurship.

Richa Sharma is a researcher with a Ph.D. in Computer Science and Engineering from PES University, Bangalore, India, complemented by an M.Tech. in Computer Science from Lovely Professional University, Phagwara, and an MCA from PCTE Group of Institutes, Ludhiana. As an Associate Professor in the Department of Computer Science Engineering at PES University, Bengaluru, Dr. Sharma brings nine years of teaching experience and five years of dedicated research expertise. Her research focuses on Natural Language Processing, Machine Learning, Deep Learning, and Generative AI. She has contributed to the field through over 16 research papers, presented at esteemed international conferences and published in reputable journals.

Chinmayi Shetty is a researcher with a BTech in Machine Intelligence and Data Science from PES University. Her work explores Natural Language Processing (NLP), machine intelligence, and backend engineering, underpinned by a strong analytical approach to understanding complex systems. She has published research on Online Hate Speech Detection, utilizing sentiment analysis and sarcasm detection techniques to enhance classification accuracy. Currently, Chinmayi is a backend developer at SuccessFactors, SAP Labs, where she develops AI-powered applications that optimize and transform business processes, driving innovation, scalability, and efficiency across a range of practical scenarios.

J. Sudarvel is an Assistant Professor of Management at Karpagam Academy of Higher Education in Coimbatore, bringing nine years of teaching experience and six years of research expertise. His academic focus spans finance and marketing, and he has supervised four Ph.D. scholars, with an additional four scholars currently under his mentorship. Dr. Sudarvel has an extensive publication record, with 75 articles to his credit, including 16 in Scopus-indexed journals, 13 in the Web of Science, and 12 in edited volumes, along with 36 presentations at international conferences. He serves as a Doctoral Committee Expert at Periyar University and holds positions on the Boards of Studies for VLB College and Nallamuthu Gounder Mahalingam College. His contributions have been recognized with awards such as Best Paper Presenter and Best Faculty, and he regularly leads workshops and delivers guest lectures on research methodology.

Bulcsú Székely has been working as a researcher at LUT University, formerly known as Lappeenranta University of Technology since 2007. His interests are logistics IT systems. He has a demonstrated history of working with organizations such as the Turku Chamber of Commerce, the University of Vaasa, and Haaga-Helia University of Applied Sciences Ltd. Dr. Székely is a member of The International Society for Professional Innovation Management (ISPIM). He has published one book, and two scientific peer-reviewed articles as a single author and co-authored several other articles as well as conference papers.

Ravi Thirumalaisamy is the Program Director for Finance, Accounting, and Economics at the Modern College of Business and Science in the Sultanate of Oman. With a Doctorate in Corporate Accounting and Finance, his research expertise spans

Corporate Accounting, Taxation, Corporate Finance, International Finance, and Security Analysis and Portfolio Management (SAPM). Dr. Thirumalaisamy brings 26 years of academic experience, coupled with a strong background in managerial roles, specializing in Quality Assurance and Program and Institutional Accreditation. Currently, he plays a key role in preparing the Business Department for AACSB Accreditation. His research, published in leading journals including SCOPUS, ISI, and Web of Science, covers areas such as corporate accounting, finance, investments, and international finance. A frequent presenter at international conferences, Dr. Thirumalaisamy also serves as a Program Reviewer for the Ministry of Higher Education, Research and Innovation in Oman.

Dipesh Vaya brings over 13 years of academic experience, marked by contributions in teaching, research, and leadership. He has published extensively in reputable international and national journals and authored several book chapters indexed in Web of Science and Scopus, published by prominent publishers such as Taylor & Francis, IEEE, Springer, and De Gruyter. His research interests span Cloud Computing, Cryptography, Artificial Intelligence, Machine Learning, and Data Analytics.

Alexey Vedishchev is an experienced IT professional and educator with over 12 years of expertise in computer science education. He holds an MSc in Information Technology and Business from Altay State Technical University in Russia, where he also completed specialized training in information management for higher education. Since 2014, he has been an instructor at the American University of Nigeria, where he teaches various programming, database, and software development courses. Alexey's teaching style emphasizes research, project-based learning, and diverse instructional methods. His professional background includes roles in IT support and management at organizations like Russian Post and Quanteq LTD, focusing on product support, IT policy development, and infrastructure planning.

R. Velmurugan is an Associate Professor of Commerce at Karpagam Academy of Higher Education, Coimbatore. With 18 years of teaching experience and 15 years in research, Dr. Velmurugan has successfully supervised 22 Ph.D. scholars and five M.Phil. scholars. He has an impressive publication record, with 152 articles in various international journals, including 54 articles in Scopus-indexed journals, 15 in Web of Science, and 76 in edited volumes. Additionally, he has presented 27 papers at international seminars and conferences, as well as 54 at national events. Dr. Velmurugan has served as an external examiner for numerous M.Phil. and Ph.D. viva voce examinations, further contributing his expertise to academic research and evaluation.

Manas Kumar Yogi is an Assistant Professor in the Department of Computer Science and Engineering at Pragati Engineering College (Autonomous), Surampalem, with over 14 years of teaching experience. He has an extensive publication record, with over 245 research papers published in the last 12 years, alongside 15 book chapters, four patents, and three authored books. His research interests encompass cybersecurity, soft computing, cyber-physical systems, the Internet of Things (IoT), and machine learning.

Foreword

In today's digital era, knowledge is one of the most valuable assets an organization can possess. The growth of digital systems, data analytics, and global connectivity has transformed how knowledge is managed, shared, and secured. Yet, with these advancements comes an unprecedented level of risk. Cyberthreats have become more sophisticated, targeting not only sensitive data but the very structures and processes that enable knowledge management. *Cybersecurity in Knowledge Management: Cyberthreats and Solutions* is a timely and essential contribution to this evolving field, providing readers with a comprehensive exploration of cybersecurity practices tailored to safeguarding knowledge assets across diverse industries.

This book examines the intersections between knowledge management and cybersecurity, addressing the unique challenges organizations face in protecting their intellectual and informational capital. Traditionally, cybersecurity efforts have focused on securing data or individual systems; however, the security needs for knowledge management go far beyond these limitations. Knowledge systems encompass a variety of functions, from content creation and collaborative sharing to data integration and organizational learning. These systems require not only robust technological defenses but also policies and practices that anticipate and mitigate potential threats. A notable strength of this book is its multidisciplinary approach. The authors bring perspectives from artificial intelligence, machine learning, information systems, and organizational behavior to create a holistic understanding of cybersecurity in knowledge management. By examining how AI and machine learning can aid in real-time threat detection, the book underscores the proactive measures that modern cybersecurity demands. The integration of these advanced technologies is crucial, as they allow organizations to anticipate cyber risks and react swiftly, thus minimizing potential disruptions to knowledge management processes.

Cybersecurity in Knowledge Management doesn't shy away from discussing the human element of cybersecurity. Social engineering, insider threats, and organizational culture play significant roles in cybersecurity, especially within knowledge-driven environments. This book thoughtfully explores how employee awareness, cultural practices, and leadership can either enhance or hinder cybersecurity efforts. Recognizing that technology alone cannot secure knowledge systems, the authors emphasize the importance of fostering a security-conscious culture and building resilience through education and employee engagement. The book also provides a series of case studies, presenting real-world scenarios that illustrate the devastating impact of cyber breaches on knowledge systems in industries like digital marketing, supply chains, and academia. Each case study is carefully chosen to highlight the specific challenges and lessons learned from actual incidents, demonstrating the immediate and long-term implications of cybersecurity lapses. By analyzing these cases, readers gain insights into the complexities of cybersecurity, as well as the critical need for responsive and adaptive security measures.

Another essential component of this book is its exploration of regulatory compliance and ethics in cybersecurity. With an increasing number of data privacy regulations, such as GDPR and CCPA, organizations must align their cybersecurity practices with legal and ethical standards. The authors provide valuable insights on how compliance can be integrated into cybersecurity strategies, helping organizations meet regulatory requirements while still achieving efficient knowledge management. The emphasis on ethical considerations also reminds readers that cybersecurity is not just about preventing attacks but ensuring responsible and respectful handling of knowledge and information. A unique feature of this book is its focus on sector-specific cybersecurity needs, particularly in the African digital landscape, supply chains, and post-COVID educational institutions. These targeted analyses bring to light the varied and often underrepresented challenges in securing knowledge systems within specific sectors. The African case, for instance, reveals the potential of artificial intelligence to counter regional cyberthreats, while the education sector analysis discusses the implications of hybrid learning models for knowledge transfer and cybersecurity in a post-pandemic world. These nuanced discussions highlight the book's commitment to providing practical solutions that cater to different industries' specific cybersecurity needs.

The book's innovative ranking metric for malware classification and detailed examination of advanced cybersecurity technologies, such as blockchain and quantum computing, set it apart from traditional cybersecurity literature. By introducing new metrics and models, the authors encourage readers to think critically about evaluating cybersecurity tools and adapting them to knowledge management systems. The inclusion of blockchain technology for secure knowledge sharing, and quantum-resistant encryption strategies, offers forward-thinking solutions that prepare organizations for the future of cybersecurity. For decision-makers, researchers, and cybersecurity professionals, *Cybersecurity in Knowledge Management* is an invaluable resource. It combines theory with actionable insights, allowing readers to apply the discussed strategies within their own organizations. Policymakers will also find this book useful for its comprehensive overview of regulatory challenges and ethical considerations in cybersecurity. With practical recommendations on adopting zero-trust architectures, privacy-enhancing technologies, and cybersecurity automation, the book provides a roadmap for both current and future cybersecurity needs.

In a world where cyber threats evolve as rapidly as technological advancements, this book is not only a guide but a call to action. *Cybersecurity in Knowledge Management* reminds us that securing knowledge systems requires constant vigilance, innovation, and adaptability. By equipping readers with the tools, insights, and understanding necessary to face these challenges head-on, this book ensures that knowledge management remains a source of strength and resilience in the digital age. I am confident that this book will serve as an essential resource for anyone involved in the fields of cybersecurity, knowledge management, and digital transformation. The authors have crafted a work that is as practical as it is thought-provoking, offering readers the knowledge needed to safeguard the informational assets that drive success and innovation in today's interconnected world.

Behrouz Aslani
University of La Verne, La Verne, California, USA

Behrouz Aslani holds a Ph.D. in Management Science and Engineering from Stanford University, where he also earned a Master's in the same field and another in Economics. With extensive global experience, Dr. Aslani has managed multinational projects in over 15 countries and provided consulting services for various sectors. He has taught a wide range of business and management courses to over 8000 students across multiple institutions, including the American University of Nigeria and California State Polytechnic University. Dr. Aslani is also the founder and president of Integrated Renewable Energy Inc., generating significant advancements in solar energy applications.

Preface

INTRODUCTION

The rapid rise of digital transformation has reshaped the way organizations manage, share, and protect knowledge. As businesses and institutions increasingly rely on digital platforms, they face significant cybersecurity threats that can compromise the integrity of their most valuable assets—knowledge and information. *Cybersecurity in Knowledge Management: Cyberthreats and Solutions* addresses this pressing need by exploring how organizations can fortify their knowledge management systems against evolving cyber threats. This book provides a comprehensive analysis of current and emerging cybersecurity practices tailored specifically for the complexities of knowledge management across various sectors, including digital marketing, education, and supply chains.

A distinguishing feature of this book is its multidisciplinary approach, combining insights from artificial intelligence, machine learning, human behavior, and regulatory compliance to present a well-rounded understanding of knowledge management security. Each chapter examines a unique aspect of cybersecurity, ranging from AI-driven threat detection to the impact of organizational culture on knowledge protection. Real-world case studies and cutting-edge research provide actionable insights and showcase how emerging technologies like blockchain, and quantum computing are shaping the future of cybersecurity in knowledge management.

This book is designed for professionals, researchers, and policymakers aiming to deepen their understanding of cybersecurity within the context of knowledge management. By offering both theoretical perspectives and practical guidance, *Cybersecurity in Knowledge Management: Cyberthreats and Solutions* equips readers with the tools needed to secure knowledge assets effectively, ensuring resilience in an ever-evolving digital landscape.

CHAPTER OVERVIEW

Authored by Rahul Dasari, Narasimha Rao Vajjhala, and Mathias Mbu Fonkam, this chapter, *The Role of AI and Machine Learning in KM and Cybersecurity*, introduces the transformative impact of Artificial Intelligence (AI) and Machine Learning (ML) in enhancing knowledge management (KM) and cybersecurity practices. The authors provide foundational insights into these technologies and their applications in automating processes like knowledge discovery, categorization, and distribution. Moving beyond KM, the chapter discusses AI and ML's vital contributions to cybersecurity, particularly in real-time threat detection, analysis, and response. With practical guidance on integrating AI/ML into KM, this chapter addresses challenges such as data governance and model training, supporting policymakers and professionals in adapting to future trends.

In this chapter, "Critically Analyzing Cybercrime Resilience Problems in the European Supply Chain Industry: A Preliminary Best Practice Radial Model", Kenneth David Strang and Bulcsú Székely explore the resilience issues European supply chains face in light of rising cybercrime. Employing machine learning techniques like heat maps and keyword synthesis, the authors provide a thematic analysis of existing literature to uncover key vulnerabilities. Their custom radial model offers a comparative analysis of best practices and current compliance levels within the industry. This research reveals that organizations should prioritize policy improvement, specialized training, and enhanced technological investment to mitigate cyber risks and foster resilience in Europe's supply chains.

J. Sudarvel, R. Velmurugan, S. Senthilkumar, Ravi Thirumalaisamy, and R. Sankar Ganesh collaborate in "Supply Chain Vulnerabilities in Knowledge Systems" to examine inherent vulnerabilities within global supply chain knowledge systems. The chapter highlights risks posed by cybersecurity threats, geopolitical dynamics, and regulatory demands. The authors discuss the importance of robust cybersecurity measures, supply chain diversification, and agile technology adoption to counter these challenges. Through case studies, they advocate for regulatory compliance and cross-industry collaboration, suggesting these strategies as foundational to building resilience in the competitive global marketplace.

In "Strategies for Secure Knowledge Management: An In-Depth Analysis of Educational Data Protection Trends and Themes", Eriona Çela, Alexey Vedishchev, and Narasimha Rao Vajjhala explore data protection strategies within educational settings, focusing on themes like data privacy, technological solutions, and risk assessment. Using a mixed-methods approach, they analyze student data privacy policies, encryption practices, and ethical data management in knowledge systems. The authors conclude with recommendations for future research, including the integration of emerging technologies such as federated learning and blockchain, positioning ethical governance as essential for secure knowledge management in education.

Ritu Pareek's chapter, "Cybersecurity: Trends and Challenges in Academia", provides an in-depth overview of cybersecurity trends and challenges specific to academic institutions. This chapter highlights the critical need for cybersecurity frameworks in protecting sensitive data within educational environments. Pareek presents case studies and best practices to emphasize an institution-wide approach to cybersecurity, involving IT professionals, faculty, and students alike. Her recommendations underscore the role of shared responsibility in fostering a secure, resilient academic environment.

Lukman Raimi a examines notable cases of cybersecurity breaches and insider threats impacting entrepreneurship in "Cases of Real-World Cybersecurity Breaches and Insider Threats to Entrepreneurship: Exploring Implications for Knowledge Management". Analyzing incidents like the Equifax and Juspay breaches, Raimi demonstrates the importance of regulatory frameworks, knowledge-sharing initiatives, and cybersecurity education within entrepreneurial ecosystems. This chapter underscores the need for global cooperation to mitigate cybersecurity risks and protect the knowledge assets of emerging enterprises.

In "The Language of Hate: An Exploration of Cyberbullying and Hate Speech Using Artificial Intelligence Techniques", authors Swati Maste, Chinmayi Shetty, Pallavi Prabhu, Richa Sharma, and Arti Arya address the challenges of cyberbullying and hate speech on social platforms and their impact on digital knowledge-sharing environments. By leveraging AI-driven techniques, they provide solutions for detecting and mitigating hate speech to foster safer online spaces. This chapter emphasizes the role of cybersecurity measures in creating inclusive digital environments, ultimately promoting healthier knowledge-sharing communities.

In "Combatting Cyberthreats in African Digital Space with Artificial Intelligence", Basil Osayin Daudu and Goddy Uwa Osimen focus on cyberthreats within the African digital space, including issues like internet fraud and cyberbullying. They examine how artificial intelligence can be deployed as a tool to counter these threats and support digital security across the continent. Their analysis places Africa within the global digital landscape, offering strategic insights into strengthening cybersecurity through AI-driven solutions.

M. V. Sujan Kumar, Ganesh Khekare, and Anurup Sankriti explore fake news detection in "Enhancing Knowledge Management Integrity through Fake News Detection: A Hybrid LSTM-XGBoost Approach for Cybersecurity", using a hybrid LSTM-XGBoost model to enhance information integrity. Their approach provides effective tools for combating misinformation, preserving trust in digital environments, and highlighting the value of advanced machine learning techniques in securing knowledge management systems.

In "An In-Depth Comparative Analysis of Malware Classification Techniques", Dipesh Vaya and Rakesh Kumar Saxena conduct a comparative analysis of malware classification models. They evaluate each on precision, accuracy, and scalability, introducing a novel ranking metric for a balanced assessment. This chapter serves as a practical guide for selecting robust classification techniques, bolstering cybersecurity in knowledge management systems.

Manas Kumar Yogi and Dwarampudi Aiswarya's chapter, "Safeguarding Knowledge Assets: Exploring the Nexus of Human Factors, Organizational Culture, and Social Engineering in KM Security", examines the influence of human factors, organizational culture, and social engineering on securing knowledge assets. They emphasize employee awareness, training, and a security-conscious culture as key defenses against cyber threats, especially in knowledge management.

Authored by Rajasekhara Mouly Potluri, Mukashev Erzhan Bulatovich, and Lohith Sekhar Potluri, "The Role of Knowledge Management in Cybersecurity for Digital Marketing Campaigns" addresses the integration of knowledge management in cybersecurity to protect digital marketing assets. The chapter examines frameworks to secure systems and applications, emphasizing the unique cybersecurity needs of digital marketing and the importance of strategic knowledge protection.

In "Knowledge Management and Cybersecurity in Higher Educational Institutions After Covid-19: A Case of Kazakhstan", Rajasekhara Mouly Potluri, Turarbekova Kuralay and Junnussov Aruan explores the post-COVID impact on knowledge transfer in Kazakhstani higher education. Through a study of 300 students, he assesses the efficacy of in-person, online, and hybrid formats while analyzing factors such as internet access and student-teacher relationships. Potluri offers recommendations for institutions adapting to the evolving digital landscape.

"Future Directions and Emerging Trends in Cybersecurity for Knowledge Management", authored by Philip Eappen, Adeyemi Abel Ajibesin, and Narasimha Rao Vajjhala, reviews advanced approaches like AI, blockchain, and quantum computing in enhancing KM cybersecurity. Emphasizing Zero Trust Architecture and privacy-enhancing technologies, the authors advocate for proactive, ethical innovation to secure knowledge assets in an evolving cyber landscape.

Cybersecurity in Knowledge Management: Cyberthreats and Solutions offers a comprehensive examination of how organizations can secure their knowledge assets in an increasingly complex digital landscape. Through the collaborative insights of experts in AI, cybersecurity, and knowledge management, this book addresses a wide spectrum of issues, from cybercrime resilience in supply chains to the impact of cyberthreats on educational institutions. Each chapter combines theoretical perspectives with practical applications, equipping readers with actionable strategies and the knowledge necessary to strengthen cybersecurity within various knowledge-driven sectors. The multidisciplinary approach provides a unique framework for understanding both the opportunities and the challenges in integrating cybersecurity with knowledge management systems.

1 The Role of AI and Machine Learning in KM and Cybersecurity

Rahul Dasari, Narasimha Rao Vajjhala, and Mathias Mbu Fonkam

1.1 INTRODUCTION

The integration of Artificial Intelligence (AI) and Machine Learning (ML) into cybersecurity represents a transformative shift in how organizations approach the protection of their digital assets. As cyber threats become increasingly sophisticated and pervasive, traditional security measures are proving inadequate. AI technologies offer advanced capabilities for threat detection, prevention, and response, thereby enhancing the overall cybersecurity posture of organizations. This chapter synthesizes recent literature on the role of AI and ML in cybersecurity, highlighting their applications, benefits, challenges, and future directions. The rise of cyber threats necessitates innovative solutions that can adapt to the evolving landscape of digital security. AI has emerged as a pivotal technology in this context, enabling organizations to analyze vast amounts of data and identify patterns indicative of potential threats. For instance, AI-driven systems can leverage ML algorithms to detect anomalies in network traffic, which may signify a cyberattack in progress (Rizvi, 2023). These systems are particularly effective in identifying new and emerging threats that traditional signature-based detection methods often overlook (Rangaraju, 2023). The ability of AI to process and analyze data at a scale allows for real-time threat detection and response, which is critical in mitigating the impact of cyber incidents (Sarker, 2022).

Moreover, AI enhances the predictive capabilities of cybersecurity systems, allowing organizations to anticipate potential attacks before they occur. By analyzing historical data and identifying trends, AI can forecast future vulnerabilities and recommend proactive measures to mitigate risks (Yang, 2022). This predictive aspect is crucial in a landscape where cybercriminals continuously evolve their tactics to exploit weaknesses in security systems. The integration of AI into cybersecurity frameworks not only improves detection rates but also reduces the time required to respond to incidents, thereby minimizing potential damage (Zeadally et al., 2020). The application of AI in cybersecurity extends beyond threat detection to include automated response mechanisms. AI systems can be programmed to initiate predefined responses to certain types of threats, significantly reducing the response time

DOI: 10.1201/9781003498094-1

and allowing human analysts to focus on more complex issues (Bresniker et al., 2019). For example, AI can automatically isolate affected systems during a detected breach, preventing further spread of the attack (Sarker, 2022). This automation is particularly valuable in environments where speed is essential, such as financial institutions and critical infrastructure sectors (Stevens, 2020).

However, the implementation of AI in cybersecurity is not without challenges. One significant concern is the potential for AI systems to be manipulated by adversaries. Cybercriminals can exploit vulnerabilities in AI algorithms, leading to false positives or negatives in threat detection (Shete, 2023). Additionally, the lack of transparency in AI decision-making processes raises ethical concerns regarding accountability and trust in automated systems (Shaukat et al., 2020). As AI technologies become more prevalent, it is imperative for organizations to address these challenges through robust governance frameworks and continuous monitoring of AI performance (Samtani et al., 2020). Furthermore, the intersection of AI and cybersecurity presents opportunities for enhancing Knowledge Management (KM) practices within organizations. AI can facilitate the collection, analysis, and dissemination of cybersecurity intelligence, enabling organizations to build a comprehensive understanding of the threat landscape (Zhang et al., 2022). By integrating AI-driven insights into KM systems, organizations can improve their ability to share knowledge and best practices across teams, fostering a culture of continuous learning and adaptation. This synergy between AI and KM is essential for developing resilient cybersecurity strategies that can withstand evolving threats.

The role of AI in cybersecurity is also evident in its application to emerging technologies such as the Internet of Things (IoT) and cloud computing. As these technologies proliferate, they introduce new vulnerabilities that require innovative security solutions. AI can enhance the security of IoT devices by enabling real-time monitoring and anomaly detection, thereby safeguarding against potential breaches (Charmet et al., 2022). Similarly, in cloud environments, AI can help organizations manage security risks by providing insights into user behavior and identifying unusual access patterns (Wu et al., 2020). The adaptability of AI technologies makes them well-suited for addressing the unique challenges posed by these rapidly evolving domains. In addition to its technical applications, AI also plays a crucial role in shaping the strategic direction of cybersecurity initiatives. Organizations are increasingly recognizing the importance of investing in AI-driven solutions to enhance their security posture and remain competitive in the digital landscape (Taddeo et al., 2019). This strategic alignment underscores the need for cybersecurity professionals to develop expertise in AI technologies and their applications, ensuring that organizations can effectively leverage these tools to combat cyber threats (Mohamed et al., 2023).

As the field of AI in cybersecurity continues to evolve, ongoing research is essential to address the challenges and limitations associated with these technologies. Future studies should focus on developing explainable AI models that can provide insights into their decision-making processes, thereby enhancing trust and accountability (Dapel et al., 2023). Additionally, research should explore the ethical implications of AI in cybersecurity, particularly concerning privacy and data protection (Karunamurthy et al., 2023). By addressing these issues, the cybersecurity community can foster the responsible and effective use of AI technologies. The integration of AI and ML into cybersecurity represents a significant advancement in the field, offering organizations enhanced capabilities for threat detection, prevention, and response. While challenges remain, the potential benefits of AI-driven solutions are substantial, particularly in the context of emerging technologies and evolving cyber threats. As organizations continue to navigate the complexities of the digital landscape, the strategic implementation of AI in cybersecurity will be critical for ensuring robust protection against cyber risks.

1.2 REVIEW OF LITERATURE

1.2.1 Overview of Cybersecurity

Cybersecurity encompasses the protection of devices, data, and networks from cybercriminals, serving as a critical pillar in safeguarding modern society's infrastructure. In earlier times, when computer use and internet access were limited, the emphasis on cybersecurity was correspondingly low. However, as digital integration has become central to virtually all aspects of personal, professional, and public life, cybersecurity has emerged as indispensable to preventing societal and economic instability. Essential personal information, such as Social Security numbers, credit card details, and home addresses, remains vulnerable without robust cybersecurity measures, posing significant risks of theft, fraud, and data exploitation. Professionals in the cybersecurity industry are continually innovating new strategies to protect against evolving threats. Beyond preventative measures, these professionals play a key role in disaster recovery, working to mitigate the damage of successful cyberattacks and restore normalcy. To do so effectively, they must stay abreast of emerging threats across diverse sectors and possess the technical versatility to assess vulnerabilities across various devices and platforms.

This chapter provides a comprehensive examination of cybersecurity threats, including phishing, malware, and denial-of-service (DoS) attacks, with focused discussions on malware variants, ransomware, and Advanced Persistent Threats (APTs). The role of AI and ML in advancing cybersecurity will be analyzed, alongside the risks these technologies introduce. Additionally, the vulnerabilities associated with IoT, and cloud security threats will be explored. Sector-specific challenges, including those faced by financial services, healthcare, and critical infrastructure, will also be addressed. Finally, the chapter will review response and mitigation strategies, covering proactive cyber defenses, incident response planning, best practices, and the importance of collaboration and information sharing across industries. Through these discussions, this chapter aims to contribute to a deeper understanding of cybersecurity's pivotal role in safeguarding information and ensuring societal resilience in the digital age.

1.2.2 Understanding Cyber Threats

In an increasingly internet and device-dependent society, cyber threats are diverse and continually evolving. This chapter identifies and examines eight predominant cyber threats that collectively represent a significant portion of security risks. Foremost among these is malware, short for malicious software, a broad category encompassing ransomware, trojans, keyloggers, and other applications designed to harm or exploit users. Each type of malware leverages distinct methods to infiltrate systems, compromising data integrity and security (Chakkaravarthy et al., 2019). Another prominent threat is Distributed Denial of Service (DDoS) attacks, where external systems flood a network with traffic, rendering it inoperable. This chapter differentiates between DDoS and DoS attacks, where the former utilizes multiple systems to initiate attacks, and the latter originates from a single source. Phishing attacks, familiar to many through email and SMS, lure users into sharing sensitive information, with various types—such as spear phishing, whaling, smishing, and vishing—targeting users in tailored ways to exploit personal and organizational data (Alkhalil et al., 2021). The chapter further explores spoofing, where attackers masquerade as trusted sources to mislead victims and gain unauthorized access to information. The types of spoofing—domain, email, and ARP—demonstrate the varied tactics used to intercept and exploit data. Supply chain attacks, which infect applications of reputable third-party vendors, have also risen, with recent research indicating that close to 40% of cyberattacks now occur through supply chains, complicating mitigation efforts due to their indirect impact on end users (Yeboah-Ofori & Islam, 2019).

Another critical threat category includes social engineering attacks, which use psychological manipulation to deceive individuals. Methods such as pretexting, business email compromise (BEC), and honeytrap schemes highlight the sophisticated, personalized tactics that make social engineering highly effective. IoT-based attacks exploit networks of connected devices, with the proliferation of 5G expanding this threat as more devices integrate into shared systems. Additionally, DNS tunneling, a method where attackers bypass security measures to install malware, has gained traction due to the ease of access to online guides (Wang et al., 2021). Finally, AI-powered attacks, utilizing adversarial AI and deepfakes, represent an emergent threat. These tools allow cybercriminals to undermine AI and ML systems by injecting malicious data or generating realistic yet fraudulent media to disseminate misinformation. This chapter presents a comprehensive analysis of these cyber threats and discusses advanced strategies for proactive cybersecurity, from detection to response mechanisms, emphasizing the growing need for adaptive, informed approaches to counteract cyber risks in a technologically reliant society.

1.2.3 Historical Perspective

Cybersecurity, initially sparked by the Creeper virus in 1970, has evolved dramatically over the past five decades. Created by Bob Thomas as a test of computer security, the Creeper virus left a trace on every infected machine. In response, Ray Tomlinson developed the Reaper, marking the first antivirus software (Pandya & Kharote, 2024). This early interaction between viruses and antivirus laid the groundwork for commercial cybersecurity solutions in the 1980s, including notable entries like NOD Antivirus and McAfee's VirusScan in 1987. Since then, the cybersecurity landscape has continued to develop, especially in response to organizational cyber threats that surfaced in the 2000s, prompting significant legal and regulatory reforms. With digital integration expanding across industries, cybersecurity has become a foundational necessity, with projections estimating the industry will reach $345 billion by 2026 (Möller, 2023).

The emergence of AI in cybersecurity is a key focus of this chapter, examining how AI transforms both offensive and defensive cyber strategies. AI's ability to automate target profiling, rapidly gather information, and learn from iterative processes has made it a powerful tool for cybercriminals. Attacks like phishing have become increasingly sophisticated through AI, which can now craft highly realistic messages that evade traditional detection methods. Additionally, cybercriminals exploit AI vulnerabilities, such as poisoning attacks, where biased data is inserted into AI systems, and evasion attacks, which manipulate data to skew AI predictions. Speculation suggests that advanced models, like Generative Pre-trained Transformers (GPTs), may enable even more adaptive and evasive threats in the future (Bengesi et al., 2024). On the defensive side, AI and ML technologies have shown considerable promise in identifying hardware and software vulnerabilities, detecting anomalies in vast datasets, and enhancing the detection of phishing and malware threats. Yet, as these technologies advance, so too will methods to counteract them, creating a complex and evolving battle between cybersecurity advancements and AI-enabled cyber threats.

This chapter also explores IoT and cloud security as critical areas within the cybersecurity landscape. IoT devices, while integral to modern society, introduce unique vulnerabilities due to limited computational resources, often lacking firewalls or encryption. These limitations make small devices, such as smart cameras or lights, attractive targets for attackers, with outdated components further compounding their security risks. Similarly, cloud computing faces inherent risks, with human error—such as incorrect API usage or misconfiguration—posing significant security threats. Insider threats within organizations also endanger data security due to the direct access individuals may have to sensitive cloud-stored data. To counteract these vulnerabilities, cloud organizations are implementing stricter access controls and adhering to regulatory standards like the Payment Card Industry Data Security Standard (PCI DSS) and the Health Insurance Portability and Accountability Act (HIPAA). This chapter provides an in-depth exploration of these emerging trends and vulnerabilities, emphasizing the importance of a comprehensive, adaptive approach to cybersecurity in safeguarding data across an interconnected digital landscape.

1.3 INTEGRATION OF AI AND ML INTO KM PRACTICES

The integration of AI and ML into KM practices represents a transformative shift in how organizations manage and leverage their knowledge assets. This integration facilitates the automation and enhancement of knowledge discovery, categorization, and distribution, thereby streamlining processes and improving decision-making capabilities across various sectors. The following discussion synthesizes insights from multiple studies to elucidate the multifaceted roles of AI and ML in KM practices. AI and ML technologies significantly enhance knowledge discovery by enabling organizations to analyze vast amounts of data efficiently. For instance, Bappy highlights that AI and ML facilitate the seamless integration of physical operations with digital intelligence, allowing for the analysis of large datasets to identify trends and predict market demands (Bappy et al., 2024). This capability is crucial for organizations aiming to extract actionable insights from their knowledge repositories. Furthermore, Singh emphasizes that AI and ML serve as catalysts for digital transformation in supply chain management, driving value creation through enhanced predictive analytics and real-time decision-making (Singh, 2023). By automating data analysis, these technologies allow organizations to uncover hidden patterns and insights that would otherwise remain dormant in traditional KM systems.

In addition to knowledge discovery, AI and ML play a pivotal role in the categorization of knowledge. The application of these technologies enables organizations to implement advanced classification algorithms that can automatically categorize information based on predefined criteria. For example, Parsaeefard et al. (2019) describe how AI can be employed in a monitoring, analysis, policy, execution, and knowledge (MAPE-K) loop, which is essential for managing internal operations and interactions with external systems. This automated categorization not only enhances the efficiency of KM processes but also ensures that relevant information is readily accessible to decision-makers. Moreover, Lainjo (2023) discusses how the integration of AI and ML into results-based management practices enhances accountability and transparency, further supporting effective categorization of knowledge. The distribution of knowledge is another area where AI and ML demonstrate significant impact. By leveraging these technologies, organizations can create intelligent systems that facilitate the timely dissemination of information to the right stakeholders. For instance, Muthaluri et al. (2024) note that AI-driven algorithms improve logistics planning and inventory management, which are critical for ensuring that knowledge is distributed effectively across supply chains. Additionally, the work of Koulaouzidis et al. (2022) illustrates how AI can enhance clinical decision support systems by providing timely and relevant information to healthcare professionals, thereby improving patient outcomes. This capability to distribute knowledge efficiently is essential for fostering collaboration and innovation within organizations.

Moreover, the integration of AI and ML in KM practices fosters a culture of continuous learning and adaptation. As organizations increasingly rely on data-driven insights, the role of human resources evolves to complement these technologies. Adeyeye emphasizes that AI and ML enable automated design optimization and predictive maintenance, which are vital for enhancing the reliability and sustainability of engineering systems (Adeyeye & Akanbi, 2024). This shift not only enhances operational efficiency but also encourages employees to engage with AI and ML tools actively, thereby promoting a culture of knowledge sharing and collaboration. However, the integration of AI and ML into KM practices is not without challenges. The evolving nature of these technologies can create uncertainty among employees regarding their roles and responsibilities. Kumar et al. (2024) highlight that technostress, stemming from the rapid advancement of AI and ML, can lead to feelings of inadequacy and knowledge gaps among workers. To mitigate these challenges, organizations must invest in training and development initiatives that empower employees to leverage AI and ML effectively in their KM efforts.

Furthermore, ethical considerations surrounding the use of AI and ML in KM practices must be addressed. The potential for bias in AI algorithms can lead to unequal access to knowledge and resources within organizations. As noted by Trim and Lee, the integration of sociocultural intelligence with AI can enhance organizational resilience and security, ensuring that knowledge distribution is equitable and just (Trim & Lee, 2022). Organizations must therefore adopt a holistic approach to AI and ML integration, balancing technological advancements with ethical considerations to foster an inclusive KM environment. The integration of AI and ML into KM practices offers substantial benefits in terms of knowledge discovery, categorization, and distribution. These technologies enable organizations to automate processes, enhance decision-making, and foster a culture of continuous learning. However, the challenges associated with technostress and ethical considerations must be carefully managed to ensure that the integration of AI and ML contributes positively to organizational KM efforts. As organizations continue to navigate the complexities of the digital landscape, the strategic implementation of AI and ML will be crucial for achieving competitive advantage and driving innovation.

1.4 APPLICATION OF AI AND ML IN CYBERSECURITY

The integration of AI and ML into cybersecurity represents a transformative shift in how organizations detect, analyze, and respond to cyber threats. This revolution is characterized by the ability of AI and ML technologies to process vast amounts of data at unprecedented speeds, enabling proactive rather than reactive security measures. Traditional cybersecurity methods, which often rely on signature-based detection, are increasingly inadequate against sophisticated cyber threats. In contrast, AI-driven systems leverage advanced algorithms to identify patterns and anomalies indicative of potential attacks, thereby significantly enhancing threat detection capabilities (Camacho, 2024). AI technologies, particularly ML algorithms, are adept at analyzing large datasets to uncover hidden threats that may not be immediately apparent. For instance, the use of deep learning techniques allows cybersecurity professionals to detect subtle indicators of compromise that traditional methods might overlook (Camacho, 2024; Rangaraju, 2023). This capability is crucial in an era where cyber threats are becoming more complex and varied, necessitating a shift from conventional security paradigms to more dynamic, AI-enhanced approaches (Sarker, 2022; Shoetan et al., 2024). The ability of AI to learn from past incidents and adapt to new threats in real time is a game-changer, providing organizations with a robust defense mechanism against evolving cyber adversaries (Camacho, 2024).

Moreover, AI's role in incident response is equally transformative. By automating routine security tasks, AI systems can significantly reduce the time it takes to respond to incidents, thereby minimizing potential damage. For example, AI can automate the identification and containment of threats, allowing human analysts to focus on more complex issues that require nuanced understanding (Adewusi et al., 2024; Camacho, 2024). This automation not only enhances efficiency but also reduces the likelihood of human error, which is a common vulnerability in cybersecurity operations (Adewusi et al., 2024). The integration of AI into Security Information and Event Management (SIEM) systems exemplifies this trend, as these systems can now leverage ML to improve threat detection and response times (Pulyala, 2023; Sarker, 2022). The application of AI in cybersecurity also extends to predictive analytics, where ML models analyze historical data to forecast potential future attacks. This proactive approach enables organizations to implement preventive measures before threats materialize, thereby enhancing overall security posture (Camacho, 2024; Shoetan et al., 2024). For instance, AI can analyze user behavior patterns to identify deviations that may indicate a security breach, allowing for timely intervention (Camacho, 2024). This predictive capability is particularly valuable in protecting critical infrastructure and sensitive data, where the cost of a breach can be substantial (Adewusi et al., 2024; Camacho, 2024).

However, the deployment of AI in cybersecurity is not without challenges. Ethical considerations, data privacy issues, and the potential for algorithmic bias are significant concerns that must be addressed as organizations increasingly rely on AI technologies (Abrahams et al., 2024; Katrakazas & Papastergiou, 2024; Udeh et al., 2024). The effectiveness of AI systems can be compromised if they are trained on biased data or if they operate without adequate oversight. Furthermore, as AI technologies evolve, so too do the tactics employed by cybercriminals, necessitating continuous adaptation and improvement of AI-driven security measures (Abrahams et al., 2024; Katrakazas & Papastergiou, 2024; Wang, 2024). The convergence of AI and cybersecurity also highlights the importance of human oversight in the security process. While AI can enhance detection and response capabilities, human analysts play a crucial role in interpreting AI-generated insights and making informed decisions based on contextual understanding (Adewusi et al., 2024; Camacho, 2024). The synergy between human expertise and AI capabilities is essential for developing a comprehensive cybersecurity strategy that can effectively address the multifaceted nature of cyber threats (Abrahams et al., 2024; Udeh et al., 2024). The application of AI and ML in cybersecurity is revolutionizing the field by enhancing threat detection, analysis, and response mechanisms. The ability to process large datasets, automate responses, and predict future threats positions AI as a critical component of modern cybersecurity strategies. However, organizations must navigate the associated challenges, including ethical considerations and the need for human oversight, to fully realize the potential of these technologies. As the cybersecurity landscape continues to evolve, the integration of AI and ML will undoubtedly play a pivotal role in shaping the future of cybersecurity practices.

1.5 RECOMMENDATIONS FOR POLICYMAKERS

Incorporating AI and ML into KM and cybersecurity strategies is becoming increasingly vital as organizations face a growing number of cyber threats. The integration of these technologies can significantly enhance the ability to manage knowledge and respond to security incidents effectively. However, it is essential to adopt best practices that ensure human oversight, address challenges related to data governance, and focus on model training to maximize the benefits of AI and ML in these domains. One of the best practices for integrating AI and ML into KM and cybersecurity strategies is to establish a robust framework for data governance. Effective data governance ensures that the data used for training AI models is accurate, relevant, and secure. This is particularly important in cybersecurity, where the quality of data directly impacts the performance of threat detection systems. Studies have shown that AI-based cybersecurity solutions can analyze vast amounts of data to identify patterns and anomalies indicative of potential threats, thus enhancing the overall security posture of organizations (Rangaraju, 2023; Rizvi, 2023; Sarker, 2022). However, without proper data governance, the risk of bias and inaccuracies in AI models increases, leading to ineffective threat detection and response (Charmet et al., 2022; Sharma & Sharma, 2024).

Moreover, human oversight remains a critical component in the deployment of AI and ML technologies in cybersecurity. While these technologies can automate many processes, the complexity and evolving nature of cyber threats necessitate human intervention to interpret results and make informed decisions (Karunamurthy et al., 2023; Taddeo et al., 2019). For instance, AI systems may identify potential threats, but human analysts are needed to assess the context and determine the appropriate response. This human-in-the-loop approach not only enhances the effectiveness of AI-driven solutions but also builds trust in the technology, as users can understand and validate the decisions made by AI systems (Karunamurthy et al., 2023; Taddeo et al., 2019). In addition to data governance and human oversight, organizations must also focus on the challenges associated with model training. The training of AI models requires large datasets that accurately reflect the types of threats an organization may face. However, obtaining such datasets can be challenging due to privacy concerns and the need for real-time data. Furthermore, the dynamic nature of cyber threats means that models must be continuously updated and retrained to remain effective. This necessitates a commitment to ongoing research and development, as well as collaboration between cybersecurity experts and AI practitioners to ensure that models are trained on the most relevant and current data (Sarker, 2022).

Looking forward, policymakers must recognize the emerging trends and potential challenges associated with the convergence of AI, ML, KM, and cybersecurity. As AI technologies continue to evolve, they will play an increasingly central role in shaping cybersecurity strategies. For instance, the integration of AI with cloud engineering is revolutionizing cybersecurity by enabling real-time threat detection and automated responses (Mohamed et al., 2023). However, this convergence also raises concerns about the reliability and accountability of AI-driven security measures, particularly considering the "black-box" nature of many AI algorithms (Zhang et al., 2022). Policymakers must therefore prioritize the development of transparent AI systems that can be audited and understood by human operators. Additionally, the rise of the IoT presents unique challenges for cybersecurity that require innovative solutions. The integration of AI and ML into IoT security frameworks is essential for addressing vulnerabilities associated with connected devices (Sarker, 2022). Standardized frameworks that outline best practices for securing IoT systems will be crucial in guiding organizations as they adopt these technologies. Policymakers should advocate for the establishment of such frameworks to ensure that organizations can effectively manage the security risks associated with IoT deployments.

As AI and ML become more prevalent in cybersecurity, there is a growing need for education and training programs that equip professionals with the skills necessary to leverage these technologies effectively. This includes not only technical skills related to AI and ML but also an understanding of the ethical implications and governance challenges associated with their use in cybersecurity (Abiodun et al., 2018; Chen et al., 2020). Policymakers should support initiatives that promote education and training in these areas to build a workforce capable of navigating the complexities of AI-driven cybersecurity. The integration of AI and ML into KM and cybersecurity strategies presents significant opportunities for enhancing organizational security and KM capabilities. However, it is essential to adopt best practices that prioritize data governance, human oversight, and effective model training. Policymakers play a crucial role in shaping the future of AI and cybersecurity by addressing emerging trends and challenges, advocating for standardized frameworks, and promoting education and training initiatives. As organizations continue to navigate the evolving landscape of cyber threats, the successful convergence of AI, ML, KM, and cybersecurity will depend on a collaborative and informed approach.

1.6 CONCLUSION

This chapter has provided a comprehensive overview of the transformative role of AI and ML in both KM and cybersecurity. AI and ML technologies have proven instrumental in enhancing the automation and efficiency of KM processes, streamlining knowledge discovery, categorization, and distribution. Their capabilities enable organizations to derive actionable insights from extensive data, facilitating informed decision-making and operational agility. AI and ML present innovative approaches to threat detection, predictive analysis, and automated response mechanisms, bolstering defenses against increasingly complex cyber threats. By leveraging AI's ability to analyze data at scale and ML's adaptive learning, organizations can anticipate and mitigate potential security risks, underscoring the necessity for these technologies in modern security strategies. However, the chapter also acknowledges the challenges of implementing AI and ML, particularly concerning data governance, ethical considerations, and the need for human oversight. Effective AI integration requires robust frameworks to address these complexities and maintain transparency, fairness, and accountability. As AI continues to evolve, its convergence with KM and cybersecurity will undoubtedly shape the future of digital security and KM, making it essential for organizations and policymakers to prioritize ethical governance and continuous learning in these domains.

REFERENCES

Abiodun, O. I., Jantan, A., Omolara, A. E., Dada, K. V., Mohamed, N. A., & Arshad, H. (2018). State-of-the-art in artificial neural network applications: A survey. *Heliyon*, *4*(11), 23–29.

Abrahams, T. O., Ewuga, S. K., Dawodu, S. O., Adegbite, A. O., & Hassan, A. O. (2024). A review of cybersecurity strategies in modern organizations: examining the evolution and effectiveness of cybersecurity measures for data protection. *Computer Science & IT Research Journal*, *5*(1), 1–25.

Adewusi, A. O., Okoli, U. I., Olorunsogo, T., Adaga, E., Daraojimba, D. O., & Obi, O. C. (2024). Artificial intelligence in cybersecurity: Protecting national infrastructure: A USA. *World Journal of Advanced Research and Reviews*, *21*(1), 2263–2275.

Adeyeye, O. J., & Akanbi, I. (2024). Artificial intelligence for systems engineering complexity: A review on the use of AI and machine learning algorithms. *Computer Science & IT Research Journal*, *5*(4), 787–808.

Alkhalil, Z., Hewage, C., Nawaf, L., & Khan, I. (2021). Phishing attacks: A recent comprehensive study and a new anatomy. *Frontiers in Computer Science*, *3*, 563060.

Bappy, M. A., Ahmed, M., & Rauf, M. A. (2024). Exploring the integration of informed machine learning in engineering applications: A comprehensive review. *Manam and Rauf, Md Abdur, Exploring the Integration of Informed Machine Learning in Engineering Applications: A Comprehensive Review*, *10*, 31–39 (February 19, 2024).

Bengesi, S., El-Sayed, H., Sarker, M. K., Houkpati, Y., Irungu, J., & Oladunni, T. (2024). Advancements in generative AI: A comprehensive review of GANs, GPT, autoencoders, diffusion model, and transformers. *Ieee Access*, *14*, 119–127.

Bresniker, K., Gavrilovska, A., Holt, J., Milojicic, D., & Tran, T. (2019). Grand challenge: Applying artificial intelligence and machine learning to cybersecurity. *Computer*, *52*(12), 45–52.

Camacho, N. G. (2024). The role of AI in cybersecurity: Addressing threats in the digital age. *Journal of Artificial Intelligence General science (JAIGS)*, *3*(1), 143–154.

Chakkaravarthy, S. S., Sangeetha, D., & Vaidehi, V. (2019). A survey on malware analysis and mitigation techniques. *Computer Science Review*, *32*, 1–23.

Charmet, F., Tanuwidjaja, H. C., Ayoubi, S., Gimenez, P.-F., Han, Y., Jmila, H., Blanc, G., Takahashi, T., & Zhang, Z. (2022). Explainable artificial intelligence for cybersecurity: A literature survey. *Annals of Telecommunications*, *77*(11), 789–812.

Chen, S.-Y., Kuo, H.-Y., & Lee, C. (2020). Preparing society for automated vehicles: Perceptions of the importance and urgency of emerging issues of governance, regulations, and wider impacts. *Sustainability*, *12*(19), 7844.

Dapel, M. E., Asante, M., Uba, C. D., & Agyeman, M. O. (2023). Artificial intelligence techniques in cybersecurity management. Cybersecurity in the Age of Smart Societies: *Proceedings of the 14th International Conference on Global Security, Safety and Sustainability*, London, September.

Karunamurthy, A., Kiruthivasan, R., & Gauthamkrishna, S. (2023). Human-in-the-loop intelligence: Advancing AI-centric cybersecurity for the future. *Quing: International Journal of Multidisciplinary Scientific Research and Development*, *2*(3), 20–43.

Katrakazas, P., & Papastergiou, S. (2024). A stakeholder needs analysis in cybersecurity: A systemic approach to enhancing digital infrastructure resilience. *Businesses*, *4*(2), 225–240.

Koulaouzidis, G., Jadczyk, T., Iakovidis, D. K., Koulaouzidis, A., Bisnaire, M., & Charisopoulou, D. (2022). Artificial intelligence in cardiology—A narrative review of current status. *Journal of Clinical Medicine*, *11*(13), 3910.

Kumar, A., Krishnamoorthy, B., & Bhattacharyya, S. S. (2024). Machine learning and artificial intelligence-induced technostress in organizations: a study on automation-augmentation paradox with socio-technical systems as coping mechanisms. *International Journal of Organizational Analysis*, *32*(4), 681–701.

Lainjo, B. (2023). The application of artificial intelligence and machine learning to enhance results-based management. *Journal of Information Systems and Informatics*, *5*(4), 1550–1568.

Mohamed, N., El-Guindy, M., Oubelaid, A., & Khameis Almazrouei, S. (2023). Smart energy meets smart security: A comprehensive review of AI applications in cybersecurity for renewable energy systems. *International Journal of Electrical and Electronics Research*, *11*(3), 728–732.

Möller, D. P. (2023). Cybersecurity in digital transformation. In *Guide to cybersecurity in digital transformation: Trends, methods, technologies, applications and best practices* Dietmar Moller (ed.) (pp. 1–70). Springer.

Muthaluri, J., Sivakumar, T., Samanta, K., & Sanne, S. H. V. (2024). Optimizing supply chain management and marketing strategies: AI and ML integration for competitive advantage. *Educational Administration: Theory and Practice*, *30*(5), 8990–8997.

Pandya, J. J., & Kharote, P. (2024). Review on cybersecurity and techniques. *International Conference on Smart Computing and Communication.*

Parsaeefard, S., Tabrizian, I., & Leon-Garcia, A. (2019). Artificial intelligence as a service (AI-aaS) on software-defined infrastructure. *2019 IEEE conference on standards for communications and networking (CSCN).*

Pulyala, S. R. (2023). The future of SIEM in a machine learning-driven cybersecurity landscape. *Turkish Journal of Computer and Mathematics Education (TURCOMAT)*, *14*(03), 1309–1314.

Rangaraju, S. (2023). AI sentry: Reinventing cybersecurity through intelligent threat detection. *EPH-International Journal of Science And Engineering*, *9*(3), 30–35.
Rizvi, M. (2023). Enhancing cybersecurity: The power of artificial intelligence in threat detection and prevention. *International Journal of Advanced Engineering Research and Science*, *10*(5), 23–37.
Samtani, S., Kantarcioglu, M., & Chen, H. (2020). Trailblazing the artificial intelligence for cybersecurity discipline: A multi-disciplinary research roadmap. *ACM Transactions on Management Information Systems (TMIS)*, *11*, 1–19.
Sarker, I. H. (2022). AI-based modeling: techniques, applications and research issues towards automation, intelligent and smart systems. *SN Computer Science*, *3*(2), 158.
Sharma, T., & Sharma, P. (2024). AI-based cybersecurity threat detection and prevention. In Perspectives on artificial intelligence in times of turbulence: Theoretical background to applications Nuno Geada, George Leal Jamil (eds.) (pp. 81–98). IGI Global.
Shaukat, K., Luo, S., Varadharajan, V., Hameed, I. A., & Xu, M. (2020). A survey on machine learning techniques for cyber security in the last decade. *IEEE Access*, *8*, 222310–222354.
Shete, S. (2023). AI in cybersecurity and user interface design beyond chatbots. *Journal of Artificial Intelligence & Cloud Computing*, *164*, 2–4. https://doi.org/10.47363/JAICC/2023
Shoetan, P. O., Amoo, O. O., Okafor, E. S., & Olorunfemi, O. L. (2024). Synthesizing AI'S impact on cybersecurity in telecommunications: A conceptual framework. *Computer Science & IT Research Journal*, *5*(3), 594–605.
Singh, P. (2023). Digital transformation in supply chain management: Artificial Intelligence (AI) and machine learning (ML) as catalysts for value creation. *International Journal of Supply Chain Management*, *12*(6), 57–63.
Stevens, T. (2020). Knowledge in the grey zone: AI and cybersecurity. *Digital War*, *1*(1), 164–170.
Taddeo, M., McCutcheon, T., & Floridi, L. (2019). Trusting artificial intelligence in cybersecurity is a double-edged sword. *Nature Machine Intelligence*, *1*(12), 557–560.
Trim, P. R., & Lee, Y.-I. (2022). Combining sociocultural intelligence with Artificial Intelligence to increase organizational cyber security provision through enhanced resilience. *Big Data and Cognitive Computing*, *6*(4), 110.
Udeh, E. O., Amajuoyi, P., Adeusi, K. B., & Scott, A. O. (2024). The integration of artificial intelligence in cybersecurity measures for sustainable finance platforms: An analysis. *Computer Science & IT Research Journal*, *5*(6), 1221–1246.
Wang, M. (2024). Generative AI: A new challenge for cybersecurity. *Journal of Computer Science and Technology Studies*, *6*(2), 13–18.
Wang, Y., Zhou, A., Liao, S., Zheng, R., Hu, R., & Zhang, L. (2021). A comprehensive survey on DNS tunnel detection. *Computer Networks*, *197*, 108322.
Wu, H., Han, H., Wang, X., & Sun, S. (2020). Research on artificial intelligence enhancing internet of things security: A survey. *IEEE Access*, *8*, 153826–153848.
Yang, K. (2022). The future of the "Metaverse": Artificial intelligence and cybersecurity. *2022 3rd International Conference on Artificial Intelligence and Education (IC-ICAIE 2022)*.
Yeboah-Ofori, A., & Islam, S. (2019). Cyber security threat modeling for supply chain organizational environments. *Future internet*, *11*(3), 63.
Zeadally, S., Adi, E., Baig, Z., & Khan, I. A. (2020). Harnessing artificial intelligence capabilities to improve cybersecurity. *IEEE Access*, *8*, 23817–23837.
Zhang, Z., Al Hamadi, H., Damiani, E., Yeun, C. Y., & Taher, F. (2022). Explainable artificial intelligence applications in cyber security: State-of-the-art in research. *IEEE Access*, *10*, 93104–93139.

2 Critically Analyzing Cybercrime Resilience Problems in the European Supply Chain Industry

A Preliminary Best Practice Radial Model

Kenneth David Strang and Bulcsú Székely

2.1 INTRODUCTION

Most people depend on the supply chain industry for survival because few societies produce all the goods and services needed for self-sufficiency. Cooperation in the supply chain industry has enabled countries to prosper through trading system integration underpinned by technology. However, cybercriminals have been taking advantage of weaknesses in supply chain integration, either for financial gains or as part of politically motivated terrorism strategies. For example, an employee of a multinational conglomerate in Hong Kong, China, transferred $25.6 million to several bank accounts managed by cybercriminals after being convinced it was authorized during a Zoom call with his stakeholders—the problem was all the other meeting attendees were deepfakes programmed using artificial intelligence and identity theft (Magramo 2024). Political warfare has spurned severe cybercrime attacks against European supply chain industry practitioners, particularly in Finland.

Supply chain organizations in Finland have experienced devastating cyberattacks that were traced back to Russia (Strang & Vajjhala, 2023, ch. 1). The Russia-Ukraine battle is a critical factor making cybercrime worse in Finland which shares an approximate 3100-km border with Russia. Finland's recent NATO membership, which was triggered by the war in Ukraine, has painted a cyber bullseye on its back for Moscow to target. Russia has perpetrated large-scale supply chain cyber-sabotage operations on critical knowledge system infrastructures in the EU (Finnish Security and Intelligence Service 2024; Elo 2024; Honkala et al., 2024). It is difficult for Finland to mitigate against these supply chain cyberattacks because there are many

DOI: 10.1201/9781003498094-2

skillful IT specialists originally from Russia's Red Army situated within Scandinavia who can be quickly transformed into cyber attackers when ordered by Moscow.

Recently, the City of Helsinki was attacked by Russia, which is considered one of the most severe cyberattacks yet (Yle News, 2024; Arghire 2024). Approximately 40,000 Helsinki city employees and numerous clients were adversely impacted. The cybercriminals stole sensitive supply chain datasets of the employees and clients. The victims are tense and unsure whether to expect ransom demands or to experience financial bankruptcy. Many victims are concerned that the leaked private information will enable cybercriminals to steal their money or use their identity credentials to commit more crimes. The city has practically shut down what was once a robust supply chain system to mitigate further losses. The 2024 Helsinki supply chain cyberattack demonstrates the key challenge of mitigating the negative consequences of fear. Furthermore, the Helsinki supply chain cyberattack illustrates domino effect problems, such as crisis, confusion, and supply chain disruptions to upstream and downstream stakeholders.

2.1.1 Research Question

There is an increased awareness of cybercrimes and cybersecurity knowledge, so it seems perplexing that cyberattacks are expanding and growing in severity, especially in the European supply chain industry. Based on the above cybercrime attacks on the supply chain industry, particularly in Finland but also in other European regions, we developed the following research question (RQ):

- Why are cybercrimes increasing in the supply chain industry, especially in Europe, and what could mitigate those risks?

2.1.2 Definitions of Key Concepts for Supply Chain Cybersecurity Risks

In this chapter, supply chain risk refers to a certain level of vulnerability considered severe enough to be addressed because it may negatively impact victims. Vulnerability is a malicious software code, a risk factor within and beyond the supply chain (Ponemon & Gemmiti 2024; Piper 2024). The digital transformation of a supply chain describes a series of digital disruptions or events that need to be monitored and supervised. This is to avoid a disaster at a critical level of supply chain vulnerability in a digital transformation process of an organization to implement a cloud (software)—platform infrastructure (Ponemon & Gemmiti 2024; Piper 2024; Arend 2024; Vajjhala & Strang 2023).

Supply chain cyber risk crisis management plans try to ensure that cyber disasters do not occur, but when they do, there are protocols to activate to respond smoothly to the disruption(s) at hand. In this chapter, crisis management refers to a long, specific cybersecurity policy process. In contrast, disaster management is a short set of emergency mitigation procedures to minimize risks after they have occurred. Companies, organizations, and supply chain levels are the same in defining vulnerability and related measures needed in case of a large-scale crisis. Some organizations emphasize business continuity and ongoing operations more than focus energy on potential cybersecurity policy development since the latter is a lower probability and priority than making a profit. One of the Big Four global management consulting firms—KPMG—claimed that cybersecurity policy development should receive equal priority to ongoing operations because if an organization fails to create an effective strategy through their policies, then they are planning to fail (Communication from Mitushi Pitti, Managing Director Cyber-Security Services, KPMG, 2024). After all, cyberattacks are more probable than improbable, particularly for larger, more profitable companies.

2.2 LITERATURE REVIEW

The authors focused on descriptive theory articles and empirical studies about cybersecurity in the European supply chain industry to answer the RQ. The literature is organized under subtopics below, first by cybersecurity weaknesses in supply chain coordination activities and, secondly, by how practitioners have made the supply chain more resilient to cybercrime.

2.2.1 Cybersecurity Weaknesses in Supply Chain Coordination

This section focuses on the practical implications of cybersecurity weaknesses in supply chain coordination and integration activities. Modern supply chains are digital—they have been transformed from manual operations by automating and integrating people, processes, and technology. Practical models discussed in industry news and virtual workshops are a starting point for theory formation. However, theoretical models are only sometimes practical enough to help practitioners understand cybersecurity risk or to prevent cybercrimes. Empirical evidence suggests that supply chain integration may increase cybercrime risk due to logistics chain complexity and open access points (SecurityScorecard 2024; Honkala et al., 2024; Robinson & Ahluwalia 2024; Medhe 2024).

Modern integration and digitalization of supply chains have fostered an information-sharing culture. While information sharing improves supply chain integration, it can also increase cybersecurity risk. If people and processes are not aligned with organizational policies to prevent cybercrime, or if policies are insufficient, this could open the door for cybercriminals. Even if an organization's people have advanced cybersecurity knowledge, mitigation cannot be effective with the supporting technology to implement the policies. The increased integration and complexity of addressing cybersecurity while also attempting to make supply chains more efficient has caused uncertainty and confusion for practitioners. IT leaders such as Chief Information Security Officers have been forced to make difficult choices to either focus on effectiveness (profitability) or cybersecurity safety. When organizations focus more on profitability, cybercriminals, therefore, have better chances to reach their goals (Canavan & Glenn, 2024).

Silos can emerge between the upstream or downstream partners in the supply chain allowing cybercriminals the opportunity to scale up the impact of fear as a tool for managing new special operations (Pool et al. 2024; Hathaliya & Tanwar 2024; Alqudhaibi et al. 2024; Tayyab, et al. 2024). In these circumstances, the training of employees becomes paramount (The Institute of Supply Chain Management 2024; Nuseir, et al. 2024; Narasimhan, et al. 2024; Dorr & McKnight 2024; Wysopal 2024; Farooq 2024). Industrial expert panel forums (Canavan & Glenn 2024; McGuire & Panache 2024; Laskowski and Hara, 2024) confirmed the findings in our literature review: That organizations do not have adequate resources at their disposal to defend themself against cybercriminal teams. Mitigating cybersecurity risks in the supply chain will require new AI-based defense systems and better organizational policies to limit the sharing of data to only those needing it. Policies must develop a cultural awareness of cybersecurity and provide training for cybercrime mitigation. Training people, processes backed by policies, and cyber-resilient technological solutions, must be used to mitigate cybercrimes.

2.2.2 Supply Chain Cyber Resilience—Theoretical and Empirical Considerations

Supply chain information sharing culture is a key tool in measuring the extent of cyber resilience of logistics networks. Practical solutions can be aligned with regard to people, processes, and technology. Empirical studies focus on the weakest links, for example, the human element of logistics knowledge systems (European Union Agency for Cybersecurity, 2024, p. 5). Many supply chain industry companies need help empowering their partner networks with up-to-date, relevant knowledge on supply chain cyber risks. The blind spot seems to be integrating open-source code into their supply chain information systems. There are instances where partners do not even realize having been infected with software bugs in their systems because of integrating open-source code (Karimi, et al. 2024; Robinson and Ahluwalia, 2024; Ponemon & Gemmiti 2024; KPMG International 2024; CrowdStrike 2024; Piper 2024). Manual procedures are still employed to evaluate open-source dependencies by 78% out of the 1278 companies interviewed (Ponemon & Gemmiti 2024). They found that 1278 companies (52% of their sample) employed AI programs to improve efficiency, but only 32% had implemented measures to assess the vulnerability level of such codes (Ponemon & Gemmiti 2024). The cybercrime risks are constantly broadening due to the use of open software and open application programming interfaces in global software supply chains.

Pool et al. point to the lack of cooperation between the partners of supply chains as a major cause of cybersecurity risk. They predicted that cybercrimes would rise in Europe. Mitigating related supply chain risks requires an appropriate information-sharing culture and standard policies. While Hathaliya and Tanwar (2024) argue open-source software is good for the supply chain industry, they also acknowledge better cybersecurity policies and training are needed to ensure risks are closed. Since digital supply chains consist of many open-source code structures, cybercrime actors have an easy target to start with their attacks. According to Hathaliya and Tanwar (2024), cyber risk mitigation requires appropriate quality control management tools. Alqudhaibi et al. (2024) emphasized the importance of comprehending the spectrum of cyber risk. They consider that supply chain collaboration is needed to carry out adequate types of employee education and training. Tayyab et al. (2024) argued that too much attention is given to cost optimization in supply chain operations. The objectives and activities therein can create new open spaces for criminals to enter the IT systems. They assert that the cybersecurity of logistic chains should be set as a high priority for the top management, with supporting policies put in writing.

Nuseir et al. (2024) stated that practitioners should focus on understanding the process of monitoring digital transformation. Their research was focused on qualitative approaches to advance companies' capabilities to respond to emerging cyberattacks. Narasimhan et al. (2024) claimed that the evolving realities of supply chains are so complex that technology cannot be leveraged to secure operations fully. Therefore, modern supply chain practitioners need ongoing training by collaborating with academia (universities) to develop better education programs. Dorr and McKnight (2024) stated that budget control is a key concern in addressing problems of securing supply chains and preventing cybercrime. They claimed a new approach is needed to align people with processes in utilizing novel technologies. Wysopal (2024) identified another concern: the lack of cultural awareness when setting up a cybersecurity defense mechanism. He argued that organizations must understand how their supply chain software functions regarding inherent risks due to open software modules. Borwik (2024) complained that companies' cybersecurity policies were severely lacking, increasing cybercrime.

Although new cybersecurity regulatory safeguards have emerged, particularly in Europe, such as data privacy protection, many companies choose to implement only some of the regulations due to the cost, particularly when legacy supply chain systems are being used. Farooq (2024) suggested cyberattacks are increasing because the supply chains are leaving application programming interfaces open to encourage better integration. Therefore, he argued that the level of cooperation between actors in supply chains must be raised to a new efficiency level to facilitate integration and prevent cybercrimes. The Institute of Supply Chain Management (2024) claimed the underlying problem was a lack of awareness and training for employees to combat emerging cybersecurity gaps. They also asserted we need better technology to make supply chains resilient to cybercrimes.

2.3 METHODS

The authors developed a mixed method research design starting with a critical literature analysis followed by a machine learning analysis of the results to identify cybersecurity risk themes the authors then used to develop recommendations to mitigate cybercrimes. The authors applied a pragmatic ideology. The term pragmatic ideology has been used to describe the researcher's philosophy or worldview in terms of what data to collect (abstract, qualitative, quantitative, or mixed), as well as what type of techniques will be used to analyze or interpret the results to develop the scientific meaning of the results (Strang, 2015). The ideology was pragmatic since we customized and adapted techniques during the project and for this chapter. Pragmatic ideologies are well-suited for process analysis and case-based reasoning (Strang, 2015). The authors used a semi-structured critical analysis approach for the literature review. Critical analysis in literature reviews is a detailed analysis of each paper to understand its credibility, how the study was designed, what the results were, and what those results mean for the scientific research community regarding generalizations (Graf, 2015). One of the key outcomes of a critical analysis literature review is a list of important papers, with findings summarized and then interpreted by the authors using keywords or themes related to the RQ.

Machine learning was applied as a pragmatic technique to help analyze the critical analysis results and answer the RQ. Machine learning was ideal for this purpose because the algorithms are flexible, and there are appropriate routines to analyze large text datasets without any labeled fields or expectations for the record structure (Strang, 2024). We aimed to use machine learning to calculate statistical measures of how representative our selected keyword themes were for the raw data from the articles. In other words, we used machine learning to validate our interpretation of the literature keywords and themes. This was an improvement of earlier thematic analysis methodology, where the results were analyzed only by researchers without any statistical estimates of validity.

2.4 RESULTS AND DISCUSSION

2.4.1 Preliminary Literature Analysis

We re-organized the critical literature review by source and retained only the most relevant papers. We then interpreted those insights to develop thematic keyword dimensions of the problems—what the researchers found lacking in their cybersecurity studies. The results of this preliminary literature analysis are listed in Table 2.1 and discussed below.

All the studies in Table 2.1 highlighted the importance of cybersecurity in different industries, including software development, upstream supply chain (supply manufacturing), and downstream (supply chain distribution). The research included third-party breaches as well as overall supply chain risk assessments. One of the first observations emerging from data analytics was how vulnerable the stakeholders were despite being a profitable company. Each study emphasized vulnerabilities, from open-source code and third-party relationships to the butterfly effect and the dark web. For example, Ponemon and Gemmiti (2024) pointed out that approximately 52% of supply chain companies employ AI modules from open software code libraries but only 32% have implemented measures to assess the cybersecurity risks or cybercrime mitigation associated with using external code. A paradoxical observation from the literature was how cybersecurity stakeholders and victims of cybercrime were resilient to the risks and actual damages. Several studies emphasized the need for cyber resilience, including the need for security performance metrics, automated risk scoring, and trend analysis. Victims were never analyzed in the literature, but cybercrime victim damages have been shared in some publicized cases.

Overall, we observed an emphasis on communicating the importance of cybersecurity to all employees in any industry worldwide. All the studies underscored the importance of cybersecurity in their respective fields. This finding means that there is no doubt that cybersecurity is an important topic across and within pretty much all industries, disciplines, and fields. We anticipate cybersecurity to generate as much literature as we saw from the COVID-19 global pandemic a few years ago. To some

TABLE 2.1
Critical Analysis of the Supply Chain Industry Literature

Source	Key Insights	Problem
Pool et al. (2024). A systematic analysis of failures in protecting personal health data: A scoping review	The two main problem categories are ignoring IT-security guidelines (policies) and lack of awareness among healthcare professionals driven by policies. There are five key stakeholder groups involved: IT-security dept., Healthcare staff, Business Associates, Healthcare Managers, Patients.	Policies
Hathaliya & Tanwar (2024). A systematic survey on security and privacy issues of medicine supply chain: Taxonomy, framework, and research challenges	Open-source code can result in lack of data protection policies, and governance/regulation, lack of data interoperability, data consistency, data heterogeneity, and system heterogeneity. The problems with data and systems will likely lead to medicine quality management concerns. The lack of ability to control the temperature within cold chains adheres to the complexity. Ultimately this is a policy problem because it may not be a good operating procedure to use open-source code.	Policies
Alqudhaibi et al. (2024). Cybersecurity 4.0: safeguarding trust and production in the digital food industry era	A core element of enhancing security in food supply chains is the continuous training of employees. This industry is behind when it comes to investments in cybersecurity compared to finance, healthcare, and transport logistics. Stakeholders do not properly understand cyber risks. The companies that have been victims are not yet identified and there is no information on what measures these firms have taken against cyberattacks.	Training
Tayyab et al. (2024). Digital Safeguards: Navigating Cyber Threats in the Logistics Industry Framework	Interconnectedness is a double-edge sword in the logistics industry. Companies are not aware that initiatives aiming at process optimization and efficiency open the doors for cyberattacks. More training is needed.	Training
Nuseir et al. (2024). The Effect of Cyber Resilience Role in the Relationship of Intelligent Information System on the E-Supply Chain: An Empirical Evidence from the UAE Healthcare Industry	The digital transformation process of supply chains is so complex and slow that the mediating role of cyber-resilience is minimal—while there is some awareness, effective policies are missing. A qualitative approach is essential to explore more deeply the case of digitization and cyber-resilience, in order to develop more effective cybersecurity policies. This study investigated 39 hospitals.	Policies

(*Continued*)

TABLE 2.1
(Continued)

Source	Key Insights	Problem
Narasimhan et al. (2024). Software Supply Chain Resiliency at Scale	The core problem of software-based supply chains is the wide range of modules and codes employed. Industry-wide standards are missing. There is no way of compiling and automating decisions. This results in a lack of technology to mitigate against cybercrimes even when organizations have policies and trained staff in place, there is just too much complexity.	Technology
Dorr & McKnight (2024). How advanced manufacturing can improve supply chain resilience and cybersecurity,	The costs of setting up cyber-safe supply chain systems in manufacturing are beyond control. "Prioritizing cybersecurity within the prism of 'people, process and technology' from the outset of a digital transformation program enables and fosters the convergence of IT and OT." Ultimately this is a lack of technology issue.	Technology
Wysopal (2024). Rising Threat: Understanding Software Supply Chain Cyberattacks And Protecting Against Them	Companies are not aware of cybercrime severity, they lack policies, a systemic approach to identify cyber risks. The cultural awareness is not there. Software Composition Analysis is the starting point for ongoing process control. Third-party components are open gates for cybercriminals.	Policies
Borwik (2024). Rethinking supply chain resilience as cyberattacks get more disruptive	Companies may be aware of cybercrime risks but do not see where the gaps in their security infrastructure are. They must set up policies to research and monitor the opening of new gaps. Incoming new regulatory requirements are pushing the limits of budgets to the extreme.	Policies
Farooq (2024). How Leaders Can Protect Supply Chains Against Cyber Risks	Companies do not possess the know-how and resources needed to tackle emerging cyber threats. Distributed multi-tiered supply chains are the standard these days. The attack surface is widening, therefore.	Training
The Institute of Supply Chain Management (2024). The Rising Threat of Cyber Fraud in Supply Chain Management	A cyberattack epidemic is emerging, resulting in a lack of awareness and training to mitigate escalating cyber risks. A wider scale of collaboration is needed within ecosystems to foster cyber resilience and digital transformation.	Training

(*Continued*)

TABLE 2.1
(Continued)

Source	Key Insights	Problem
Canavan and Glenn (2024). Protecting against Software Supply Chain Attacks	Supply chain software consists of many different types of code. Operational urge to do more with less, in isolated teams. No visibility and no transparency. Broadening attack surface: built-in source code or package including open-source code. Problems remain hidden/ unknown vulnerabilities and compromised dependencies. Secret repository exposed. Better policies are needed to avoid using open-source code.	Policies
McGuire & Panache (2024). Deep Dive: Software Supply Chain Threats	Open-source code plays a central role in supply chain code. There are uncovered dangers lurking from within the knowledge systems. Only circa 70% of companies admit that their current solutions are inadequate. Code developers are unaware of threats. Organizations need policies to force developers to avoid using risky code.	Policies
Laskowski and Hara (2024). Breaches Beyond Borders: A Deep Dive into Third-Party Cybersecurity Breaches	The cascading effect of vulnerabilities. Threat actors can scale up their operations with minimal effort. Supply chain ecosystems are complex. Circa 98% of organizations have a relationship with a third party that has been breached—organizations need policies to limit partnering with at-risk suppliers.	Policies

degree, the pandemic stalled global cybercrimes. Still, the latter escalated during the late phase of the pandemic shutdown period when most younger people recovered but had to remain isolated. Cybercrime seemed to increase in unison with the Russian invasion of Ukraine exponentially.

We observed a common finding in key studies identifying a significant cybersecurity risk based on programming practices in industries that develop commercial software and large companies that program their systems. Specifically, we uncovered the weakness of the open-source code. This exists because talented programmers share their modules, functions, and libraries for various programming languages, but bad actors either purposefully add malevolent routines or third parties host such libraries, and they unethically embed code to leave a door open for cyber terrorists to leverage later. Two studies highlight open-source code's role in supply chain threats. Thus, we recommend that the CIO and other technology decision-makers be cautious in allowing or encouraging staff to download or use any open software libraries that are not certified by one of the larger credible vendors, including but not limited to Google, Microsoft, or IBM.

Without question, most of the researchers identified a need for improvement in cybersecurity awareness and cybercrime prevention. Most of the studies suggest that current solutions are inadequate and that there is a need for improvement. The recommended improvements ranged from increasing employee, victim, or decision-maker awareness of cybercrime vulnerabilities and, at the other end, to assessing the risks. Although no practical techniques were suggested, we can conclude from this analysis that the next step would be to explore how to increase cybercrime awareness and put safeguards into the technical environment to prevent cyberattacks—if not already done. We also observed interesting differences in how information technology and supply chain decision-makers perceive the importance of this topic in their workplace. Each study we reviewed had a specific focus, from software supply chain attacks and third-party cybersecurity breaches to supply chain risk and building blocks of cyber resilience. This illustrates what decision-makers and academic researchers think are crucial global cybersecurity topics to examine.

The core message from the above literature was that due to the complexity and slow motion of digital transformation projects, there are many blind spots within supply chain IT—ecosystems. The people, processes, and technologies involved are not aligned well enough. Companies still pursue a more with less strategic approach. New emerging regulatory requirements for knowledge systems ironically open doors for cybercriminals to attack the supply chain. The above literature illustrates an increasing feeling of panic among supply chain knowledge system operators and users. For criminal actors, it is easy to remain hidden while setting up efforts to leverage chaos.

The results of the literature reviews paint a conceptual picture of where this industry and practice are heading. The researchers suggest different solutions, from prioritizing cybersecurity and fostering transparency to developing a better academic research process and implementing critical key performance indicators (KPIs) into the organization's governance to prevent cybercrime. Subsequently, the most important result in Table 2.1 was that three key problem areas were identified by the researchers: lack of cybersecurity policies, inadequate cyberattack training, and insufficient technological resources. These problems interest stakeholders because of the increased cybercrimes, resulting in expensive financial losses and devastating humanistic damages to victims (particularly financial scams and medical record-based ransom extortions).

2.4.2 Machine Learning Content Analysis

We performed a thematic analysis using machine learning to identify, analyze, and report new patterns associated with the three problem areas: lacking policies, inadequate training, and insufficient technology. This analysis could help stakeholders understand the overarching themes across the different industries and provide the seed for additional academic studies. Figure 2.1 depicts a heat map of the keywords the cybersecurity leaders, researchers, practitioners, and executive decision-makers identified. We can observe from Figure 2.1 that verbs like additional analysis "needed" dominated the findings across all industries and countries represented in the data.

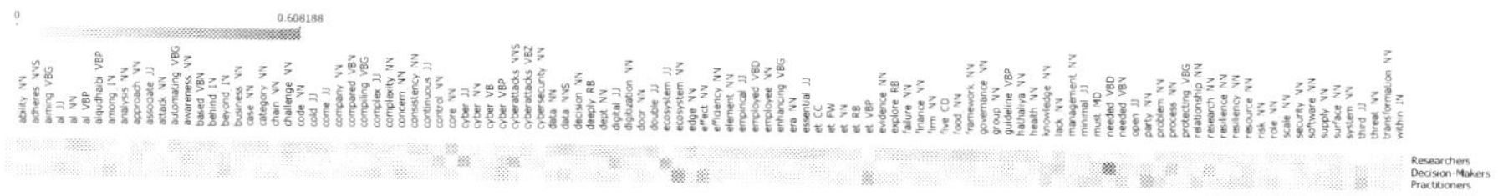

FIGURE 2.1 Heat map of ML thematic data analytics emphasizing keywords from cyberattacks.

FIGURE 2.2 Supply chain cybercrime risk keywords extracted from the critical literature analysis.

We then performed a keyword analysis of the thematic data from the heat map. The goal was to produce a visual diagram of the most critical 50 keywords, using font size to indicate the frequency or importance. The keyword diagram is more accessible to interpret than the heatmap because it has one dimension: importance (based on the size of the word). Figure 2.2 illustrates the most critical keywords extracted from the thematic heatmap data.

We may interpret the analysis in the thematic table in Figure 2.2 to answer the first part of the RQ and reveal why cybercrimes are increasing in the supply chain industry, especially in Europe. Based on the literature, cybercrimes are rising because of a strategic divide between becoming aware of a cybercrime risk and, at the other extreme, having the technology to prevent cybercrime before it occurs. We can distinguish in the content analysis research between the lack of cybercrime knowledge (usually by policy decision-makers) and the lack of cybersecurity policy awareness (by those with technological cyberattack resilience knowledge). In other words, those with problems do not connect with those who can solve the issues, and vice versa.

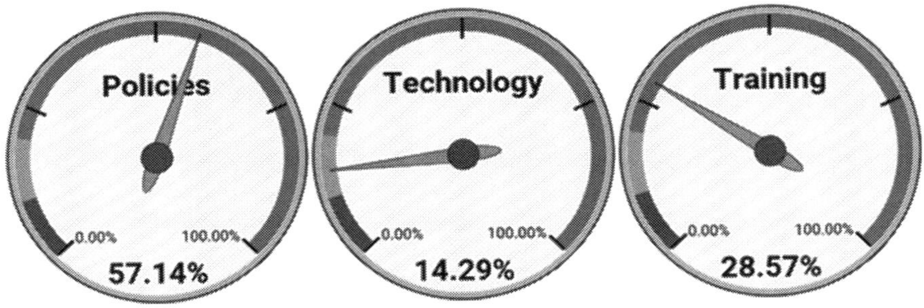

FIGURE 2.3 Key reasons for supply chain cybercrime risk—What needs to be improved.

Figure 2.3 illustrates a dashboard of the severity of cybercrime problems based on the literature critical analysis using gauges. Here, we analyzed the themes found from the critical analysis and contrasted them using counts of how many researchers had identified a specific problem to create a quantitative coefficient for each gauge. A gauge was developed for each theme to measure its severity compared to the others. This was done without statistical replacement, so each study resulted in one outcome theme. The gauge scales were percentages based on 100%, where colors were defined to highlight the severity level, with green up to 10%, orange up to 20%, and anything beyond 20% in red. More significant gauge coefficients can be interpreted as dangerous because the number identifies the percentage of studies where researchers determined that specific problem as the outcome. The other perspective of Figure 2.3 is that each gauge implies what needs to be improved and to what extent cybercrime resiliency can be achieved.

The most significant problem illustrated in Figure 2.3 was the policies gauge 57.1% (rounded). Over half of the problems identified in the literature could be reduced to the need for an effective policy addressing cybercrime risk for that organization. Organizations must have specific cybersecurity risk prevention policies, not generic wish lists. This also points to the need for organizations to conduct more research to determine if their existing policies need revision. Good examples were when software engineering companies allowed their developers to use open-source libraries, where a cybercrime may have embedded malware. Open-source libraries are safe if a credible vendor has certified them risk-free and provides rigorous authentication to ensure cybercriminals cannot change the code.

The next most severe problem in Figure 2.3 was identified by the training gauge at 28.6%, meaning that organizations needed to be made aware of the need for training or educating their staff against cybersecurity risks. Even when organizations have policies to prevent cybercrime, staff need to know how to detect risks and how to mitigate against future cybersecurity attacks. If decision-makers do not have a cybersecurity policy, they will likely not be able to assess if they have suitable training programs to avoid cybercrime attacks. Even if an organization is aware of cybercrime risk, with training, front-line and back-office staff will know how to detect or prevent it. The technology gauge was 14.3%, which refers to missing technology to address cybersecurity risks. Sometimes, the ultimate problem was the complexity of supply chain integration, where technology still needed to be created to mitigate against cybercrimes fully. The last significant observation is that all three of the above gauges were rated at the orange level, 20% or higher, signaling an emergency to act against cybercrime. An overall observation from the gauges in Figure 2.3 was that all three problems were in the moderate severity scale, over 10%, and two of the three (policies and training)—which represented approximately 86% of the problems—were well over the 20%-gauge level in the implied danger zone. Thus, these gauges imply a form of organizational maturity, starting with policies, then training, and finally having the requisite technology in place.

The problem is that decision-makers downstream in the supply chain, such as distributors, understand the importance of cybersecurity policy, but they need more accurate knowledge about preventing cybercrime risks. Technology specialists upstream in the supply chain have the competency to detect or prevent cybercrime risk. Still, upstream technology specialists need to consider downstream supply chain cybercrime policies to know when, where, why, and when cybercrime attacks may occur. This phenomenon could be interpreted as a disconnection between cybersecurity risk perception decision-makers and cybercrime technological resiliency, which makes the entire supply chain, upstream and downstream, vulnerable. This leads, in turn, to gaps in communicating critical information between the actors, creating a bullwhip effect (Haugli-Sandvik et al. 2024; KPMG International 2024; Strang 2024; Navajas-Adán et al. 2024; Vajjhala & Strang 2023). As a result, silos prevent cross-industry cyber resilience (Arroyabe et al. 2024; European Union Agency for Cybersecurity, 2024; Radanliev 2024; Strang 2024; Tzavara and Vassiliadis, 2024; Dupont et al., 2023). The key problem of observing the facts related to these two themes of insufficient coordination between cybersecurity decision-makers and cybercrime technological resilience is related to the disconnection between stakeholder perceptions (SecurityScorecard, 2024; KPMG International 2024; PwC Research 2024; Buffigton et al., 2024; Ponemon & Gemmiti 2024; Sadeghi et al. 2024; Vajjhala and Strang, 2023).

Subsequently, the solution is to connect these major stakeholders, the downstream cybercrime policy decision-makers, with the upstream cybercrime technological resilience specialists. Since these two stakeholders often exist in different parts of the supply chain or at opposite ends of large companies, executives must revise standard operating procedures and organizational structures to become key constituencies. Project management practices, including Agile Kaizen or Sprints, could be used to initiate these efforts. By their definition, these types of projects are agile and less overhead-intensive than commercial projects. A sprint project could be formed by a single executive with the power to bring together the cybercrime policy decision-maker with one or more cybercrime technological resilience specialists to synergize a better solution to cybercrime risk. Organizational change is required, as we can see from the literature review, as cybercrime can be very costly to organizations. The cost of preventing or mitigating cybercrime will be less than the cost if a cybercrime occurs. Executive decision-makers must think about that axiom. The cost of cybercrime prevention will be significantly less than that of cybercrime repair. More so, how would we know if cybercrime was repaired without the involvement of specialists who can access the cybersecurity systems and the cybercrime damages simultaneously?

Cybersecurity Awareness Frameworks and Training Models are tools for leveraging best practices and lessons learned from past cyberattacks for highly comprehensive methods of cybercrime risk governance (Taherdoost 2024; Karagiannis et al. 2024; Kouper and Stone 2024; VTT 2024; Cristiano et al., 2023). Herein there are possibilities for new interdisciplinary initiatives to close the existing gaps and silos related to emerging knowledge-sharing platforms.

2.5 CONCLUSION AND LIMITATIONS

The researchers answer the RQ in this chapter by proposing why cybercrimes were increasing in the supply chain industry, especially in Europe, and what could be done to mitigate those risks. We will explain the answer to each part of the RQ. In summary, the critical analysis of the literature pointed to one major reason for the increase of cybercrimes in Europe, particularly in the supply chain industry, warfare increased dramatically, which can be attributed to the military operations of Russia against Ukraine and the allies of Ukraine, including Finland. This is evident when examining the Russian invasion of Ukraine, which has spilled over into neighboring countries like Belarus, Poland, and especially Finland. Despite the need for further dedicated inter-continental collaboration among practitioners and research entities in the field of cyber resilience of supply chains, new supply chain knowledge gaps and silos among systems appeared. We found key problems associated with the increased cybercrimes: lack of customized organizational policies, lack of staff training to identify and mitigate against cybercrimes, and lack of technology to prevent or mitigate cyberattacks, primarily due to supply chain integration complexity.

To address the second part of the RQ answer, we propose what could be done to mitigate against the cybercrime risks we identified from the critical analysis. We developed a preliminary best practice radial model of targets for cybercrime resilience—what needs to be improved—as shown in Figure 2.4. We developed this radial model by examining the severity levels from the gauge data underlying Figure 2.3 for the three problem themes of policies, training, and technology. We took the reciprocal, by subtracting each gauge data value from 100%, which produced an estimated compliance level reading. In Figure 2.4, the estimated current compliance level for policies is 42.86%; for technology, it is 85.71%, while training is 71.43% (shown by the blue data line series). The interpretation of these values is that larger percentages are better, and close to 100% is a best practice. Thus, the largest gap in current supply chain cybercrime resiliency, based on Figure 2.4, is the lack of rigorous customized corporate cybersecurity policies, at 57%. Next is the gap of 24% in training staff for cybercrime identification, followed by implementing technology to prevent cybercrimes.

The target benchmarks in Figure 2.4 were set at 100% for policies because we argue every corporation ought to have a relevant cybersecurity policy customized to their organization—not a generic catch-all type of statement. We set the training bar at 95% because staff need to be aware and trained about how to detect cybercrime and how to mitigate against it. Of course, there could be two groups: those looking for cybercrimes and another higher-skilled team of computer scientists who can destroy malware and other cybercrime technology. We were more tolerant

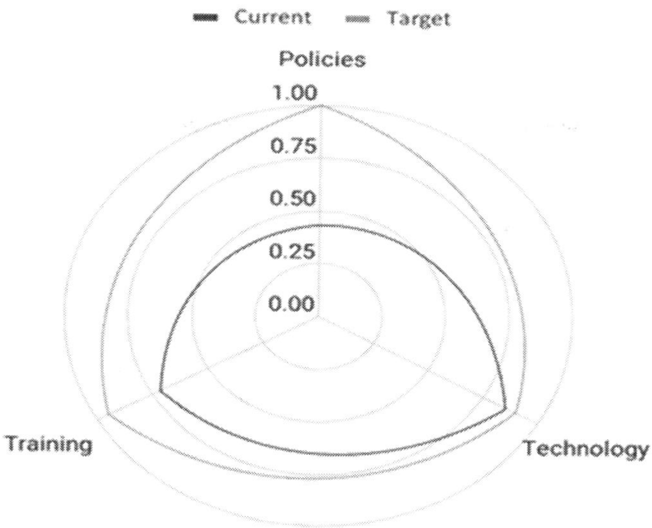

FIGURE 2.4 Radial model for supply chain cybercrime resilience—What needs to be improved.

of a lower threshold for technology, which we set at 90%, because we acknowledge the integration complexity of the existing supply chain industry, making it all but impossible to mitigate against cybercrime risk completely. Additionally, technology always advances faster than we can keep up with policies and training, so technology may never reach 100%. Nevertheless, we can see the gaps in current cybersecurity practices and where improvements are needed. We must also caution that this data covers the supply chain industry, so these recommendations may only be generalized in some sectors.

In the future, we recommend that expert knowledge system actors evaluate supply chains to identify cybersecurity risks using our radial compliance analysis. These evaluators must have a well-established overall picture of the grim situation. Research initiatives between industry sectors will yield further benefits in aligning the interpretation of expectations, what to measure, and how to stop the increasing cybercrime attack trends in the supply chain industry. Public-private partnerships are needed to stimulate a new stream of investment into generative AI-based threat detection and prevention training models. Green cybersecurity infrastructure for smart circular design and manufacturing technology could be a setting to investigate. Practical consortium project studies could bring in new empirical datasets, for example, feedback from the actors involved in the fields.

REFERENCES

Alqudhaibi, A., Krishna, A., Jagtap, S., Williams, N., Afy-Shararah, M., & Salonitis, K. (2024). Cybersecurity 4.0: Safeguarding trust and production in the digital food industry era. *Discover Food*, *4* (2). https://doi.org/10.1007/s44187-023-00071-7

Arend, C. (2024, May 23). Are your business applications holding you back on your digital business journey, International Data Corporation (IDC) Europe - webinar, [Video]. https://www.idc.com/eu/events/71525-idc-kyndryl-webinar#section_10

Arghire, I. (2024, May 14). Student, personnel information in city of Helsinki cyberattack. *Security Week.* https://www.securityweek.com/student-personnel-information-stolen-in-city-of-helsinki-cyberattack/

Arroyabe, M. F., Arranz, C. F.A., De Arroyabe, I. F., de Arroyabe, H. C. F. (2024). Revealing the realities of cybercrime in small and medium enterprises. *Understanding fear and taxonomic perspectives, Computers & Security, 141*, 103826. https://doi.org/10.1016/j.cose.2024.103826

Borwik, S. (2024, January 25). Rethinking supply chain resilience as cyber-attacks get more disruptive. *Industry News, techUK.* https://www.techuk.org/resource/rethinking-supply-chain-resilience-as-cyber-attacks-get-more-disruptive.html

Buffigton, J., Russel, D., Webb, J. (2024, January). Regional Summary from the 2024 Data Protection Trends Report, EMEA Edition, Veeam Insights, Veeam Software. https://engage-insights.com/white-paper/2024-data-protection-trends-report/download

Canavan, B. & Glenn, E. (2024, April 24). Protecting against Software Supply Chain Attacks, THALES & digicert, A BrightTALK - webinar, [Video]. https://www.brighttalk.com/webcast/2037/607859?utm_source=brighttalk-portal&utm_medium=web&utm_campaign=topic&utm_content=upcoming

Cristiano, F., Kurowska, X., Stevens, T., Marie Hurel, L., Fouad, N. S., Cavelty, M. D., Broeders, D., Liebetrau, T. & Shires, J. (2023). Cybersecurity and the politics of knowledge production: Towards a reflexive practice, *Journal of Cyber Policy*, *8*(3), 331–364. https://doi.org/10.1080/23738871.2023.2287687

CrowdStrike (2024). *Global Threat Report.* https://www.crowdstrike.com/global-threat-report/

Dorr, M. B. and McKnight, L. (2024, March 27). How advanced manufacturing can improve supply chain resilience and cybersecurity. *The World Economic Forum.* https://www.weforum.org/agenda/2024/01/advanced-manufacturing-improve-supply-chain-resilience-cybersecurity/

Dupont, B., Shearing, C., Bernier, M., Leukfeldt, R. (2023). The tensions of cyber-resilience: From sensemaking to practice, *Computers & Security, 132.* https://doi.org/10.1016/j.cose.2023.103372

Elo, V. (2024, March 14). Finland is to boost its high-technology profile in NATO – An accelerator and two test centres to be established in Finland, Press release. *Ministry of Defence.* https://valtioneuvosto.fi/en/-/finland-to-boost-its-high-technology-profile-in-nato-an-accelerator-and-two-test-centres-to-be-established-in-finland

European Union Agency for Cybersecurity. (2024, February). *Best practices for cyber crisis management*. https://doi.org/10.2824/767828. https://www.enisa.europa.eu/publications/best-practices-for-cyber-crisis-management

Farooq, B. (2024, February 12). How leaders can protect supply chains against cyber risks. *Spiceworks Inc.* https://www.spiceworks.com/supplychain/supply-chain/guest-article/how-leaders-can-protect-supply-chains-against-cyber-risks/

Finnish Security and Intelligence Service (2024, March 26). The threat of Russian intelligence and malign influence remains elevated in Finland. Press release. https://supo.fi/en/-/the-threat-of-russian-intelligence-and-malign-influence-remains-elevated-in-finland

Graf, L. (2015). Critical analysis using four case studies across industries. In *Handbook of research design in business and management* (pp. 293–318, chap. 17). Springer Palgrave Macmillan US. https://doi.org/10.1057/9781137484956

Hathaliya JJ, & Tanwar S. (2024). A systematic survey on security and privacy issues of medicine supply chain: Taxonomy, framework, and research challenges. *Security and Privacy*, e377. https://doi.org/10.1002/spy2.377

Haugli-Sandvik, M., Lund, M.S. & Bjørneseth, F.B. (2024). Maritime decision-makers and cyber security: deck officers' perception of cyber risks towards IT and OT systems. *International Journal of Information Security, 23*, 1721–1739. https://doi.org/10.1007/s10207-023-00810-y

Honkala, T., Kovanen, K., Sinkkilä, L. (2024, February). Trend report 2024: Reimagining resilience. *VTT Technical Research Centre of Finland Ltd.* https://doi.org/10.32040/2024. https://www.vttresearch.com/sites/default/files/2024-02/ENG_VTT_TrendReport2024_Final%203.pdf

Karagiannis, S., Magkos, E., Karavaras, E. A. Karnavas, M. N. Nikiforos, C. Ntantogian (2024). Towards NICE-by-design cybersecurity learning environments: A cyber range for SOC teams. *Journal of Network and Systems Management, 32*, 42. https://doi.org/10.1007/s10922-024-09816-w

Karimi, K., Campos, F., Drummond, B., Parham, G. (2024). *All supply chains are digital – Cybersecurity for a software-defined future*. Research Insights, IBM Institute for Business Value, IBM Corporation. https://www.ibm.com/downloads/cas/PNKRGA15

Kouper, I and Stone, S. (2024). Data sharing and use in cybersecurity research. *Data Science Journal, 23*(3), 1–19. https://doi.org/10.5334/dsj-2024-003

KPMG International (2024, January). Cybersecurity considerations 2024 – Technology innovations demand strategic pragmatism. https://kpmg.com/us/en/articles/2024/cybersecurity-considerations-2024.html

Laskowski, M. & Hara, P. Prudhomme (2024). Breaches Beyond Borders: A Deep Dive Into Third-Party Cybersecurity Breaches. SecurityScorecard, A BrightTALK webinar March 12

Magramo, K. (2024). British engineering giant Arup revealed $25 million deepfake scam victim. May 17, Hong Kong (CNN), Cable News Network A Warner Bros Company. https://edition.cnn.com/2024/05/16/tech/arup-deepfake-scam-loss-hong-kong-intl-hnk/index.html

McGuire, M. & Panache, D. (2024, April 20). Deep dive: Software supply chain threats, *Synopsis & Reversinglabs*, A BrightTALK - webinar, [Video]. https://www.brighttalk.com/webcast/13983/609545?utm_source=brighttalk-portal&utm_medium=web&utm_campaign=topic&utm_content=upcoming

Medhe, K. (2024, January 31). Supply Chain Risk, Cloud Security Alliance CloudBytes, A BrightTALK - webinar, [Video]. https://www.brighttalk.com/webcast/10415/602964

Narasimhan, V.L., Ramaswamy, S., & Mphale, O. (2024). Software supply chain resiliency at scale. In: *ICT: Applications and Social Interfaces* ICTCS 2023. Lecture Notes in Networks and Systems, vol *908*. Springer, Singapore. https://doi.org/10.1007/978-981-97-0210-7_37

Navajas-Adán, J., Badia-Gelabert, E., Jiménez-Saurina, L. et al. (2024). Perceptions and dilemmas around cyber-security in a Spanish research center after a cyber-attack. *International Journal of Information Security, 23*, 2315–2331. https://doi.org/10.1007/s10207-024-00847-7

Nuseir, M.T., Alquqa, E.K., Alzoubi, H.M., Alshurideh, M.T., Al Kurdi, B.& AlHamad, A. (2024). The effect of cyber resilience role in the relationship of intelligent information system on the e-supply chain: An empirical evidence from the UAE healthcare industry. In *Cyber security impact on digitalization and business intelligence*. Studies in Big Data, vol 117. Springer. https://doi.org/10.1007/978-3-031-31801-6_5

Piper, S. (2024, March 20). Cyber Exposure Management: Evolution or Revolution? A BrightTALK - webinar, [Video]. https://www.brighttalk.com/webcast/260/606421

Ponemon, L. & Gemmiti, S. (2024 May 23). By the Numbers: Software Supply Chain Security Risk, Ponemon Institute & Synopsys, a BrightTALK - webinar, [Video]. https://www.brighttalk.com/webcast/13983/611413

Pool, J., Akhlaghpour, S, Fatehi, F., Burton-Jones, A. (2024). A systematic analysis of failures in protecting personal health data: A scoping review, *International Journal of Information Management, 74*. https://doi.org/10.1016/j.ijinfomgt.2023.102719

PwC Research (2024). The C-suite playbook - Putting Security at the epicenter of innovation: Findings from 2024 Global Digital Trust Insights, Global Center of Excellence for market research and insight, PwC. https://www.pwc.com/gx/en/issues/cybersecurity/global-digital-trust-insights.html

Radanliev, P. (2024). Cyber diplomacy: defining the opportunities for cybersecurity and risks from artificial intelligence, IoT, blockchains, and quantum computing. *Journal of Cyber Security Technology*, 1–51. https://doi.org/10.1080/23742917.2024.2312671

Robinson, C. & Ahluwalia, M. (2024, March 21). The building blocks of cyber resilience, Tata Communications, A BrightTALK - webinar, [Video]. https://www.brighttalk.com/webcast/15863/609122?utm_source=brighttalk-portal&utm_medium=web&utm_content=The%20Building%20Blocks%20of%20Cyber%20Resilience&utm_term=search-result-1&utm_campaign=webcasts-search-results-feed

Sadeghi, K. R., Ojha, D., Kaur, P., Mahto, R. V., & Dhir, A. (2024). Explainable artificial intelligence and agile decision-making in supply chain cyber resilience. *Decision Support Systems*, *180*, 114194. https://doi.org/10.1016/j.dss.2024.114194

SecurityScorecard (2024, January). Cyber conflicts and the erosion of trust: Introducing the cyber resilience scorecard. *World Economic Forum Annual Meeting*. https://securityscorecard.com/reports/cyber-resilience-scorecard/

Strang, K. D. (Ed.). (2015). *Handbook of Research Design in Business and Management*. Springer Palgrave Macmillan US. https://doi.org/10.1057/9781137484956

Strang, K. D. (2024). Cybercrime risk found in employee behavior big data using semi-supervised machine learning with personality theories. *Big Data and Cognitive Computing*, *8*(4), 4. https://doi.org/10.3390/bdcc8040037

Strang, K. D. & Vajjhala, N. R. (2023). Ch. 1 Why cyberattacks disrupt society and how to mitigate risk. In *Cybersecurity for decision makers*. CRC Press Taylor & Francis. https://www.routledge.com/Cybersecurity-for-Decision-Makers/Vajjhala-Strang/p/book/9781032334974

Taherdoost, H. (2024). A critical review on cybersecurity awareness frameworks and training models, *Procedia Computer Science*, *23*, 1649–1663. https://doi.org/10.1016/j.procs.2024.04.156

Tayyab, M., Hameed, K., Jhanjhi, N. Z., Zaheer, A., & Qamar, F. (2024). Digital safeguards: Navigating cyber threats in the logistics industry framework. In N. Jhanjhi & I. Shah (Eds.), *Navigating cyber threats and cybersecurity in the logistics industry* (pp. 258–299). IGI Global. https://doi.org/10.4018/979-8-3693-3816-2.ch010

The Institute of Supply Chain Management (2024, February 26). The rising threat of cyber fraud in supply chain management. *General News*. https://www.ioscm.com/blog/the-rising-threat-of-cyber-fraud-in-supply-chain-management/

Tzavara, V., & Vassiliadis, S. (2024). Tracing the evolution of cyber resilience: A historical and conceptual review. *International Journal of Information Security*, *23*, 1695–1718. https://doi.org/10.1007/s10207-023-00811-x

Vajjhala, N.R., & Strang, K.D. (Eds.). (2023). *Cybersecurity for decision makers* (1st ed.). CRC Press. https://doi.org/10.1201/9781003319887

VTT (2024). CyberMaturity tool. *Technical Research Centre of Finland Ltd*. https://cybermaturity.vtt.fi/

Wysopal, C. (2024, February 6). Rising threat: Understanding software supply chain cyberattacks and protecting against them. Forbes Technology Council. https://www.forbes.com/sites/forbestechcouncil/2024/02/06/rising-threat-understanding-software-supply-chain-cyberattacks-and-protecting-against-them/?sh=504d663f6907

Yle News (2024, May 13). Major data breach could affect up to 120k pupils, guardians and city workers in Helsinki. Yle Broadcast Media and Distribution, Finland. https://yle.fi/a/74-20088448

3 Supply Chain Vulnerabilities in Knowledge Systems

J. Sudarvel, R. Velmurugan, S. Senthilkumar, Ravi Thirumalaisamy, and R. Sankar Ganesh

3.1 INTRODUCTION

In the contemporary global landscape, characterized by rapid technological advancements and increasing interconnectedness, the role of knowledge systems in fostering innovation, enhancing productivity, and maintaining competitive advantage across industries cannot be overstated. Knowledge systems encompass the intricate web of information, data, and intellectual assets that are pivotal to the success of organizations. From facilitating research and development to optimizing production and distribution processes, these systems are the lifeblood of modern enterprises. According to Peter Drucker, a pioneer in management theory, organizations operate within a "knowledge economy" where the effective utilization of knowledge resources is paramount for sustained success (Drucker, 1993). In this context, the efficient flow of information and intellectual capital through supply chains becomes not only advantageous but also essential. However, the reliance on interconnected networks exposes these systems to a multitude of vulnerabilities. The concept of supply chain vulnerabilities in knowledge systems encompasses a broad spectrum of challenges. These include, but are not limited to, cybersecurity threats, intellectual property risks, geopolitical instability affecting supply routes, and disruptions caused by natural disasters or global pandemics. As noted by scholars like Paul D. Cousins and Richard B. Handfield, the globalization of supply chains has intensified these risks, making enterprises more susceptible to disruptions that can ripple across their operations (Cousins & Handfield, 1996). Cybersecurity emerges as a critical concern within knowledge systems, with malicious actors targeting valuable intellectual property and sensitive data. The interconnected nature of modern supply chains amplifies these risks, as highlighted by recent incidents of cyberattacks on major corporations compromising vast amounts of proprietary information and disrupting operations globally (Rosenzweig, 2019a, 2019b).

Moreover, the protection of intellectual property (IP) poses another significant challenge. Companies invest heavily in developing and safeguarding their intellectual assets, yet vulnerabilities in supply chains can expose them to theft or unauthorized use. This issue is exacerbated in international operations where differing legal

DOI: 10.1201/9781003498094-3

frameworks and enforcement mechanisms complicate IP protection efforts (Maskus, 2000). Geopolitical factors further complicate supply chain dynamics in knowledge systems. Trade tensions, changes in regulatory environments, and political instability can disrupt supply routes or impose restrictions on the movement of critical components or information. Such uncertainties necessitate strategic foresight and contingency planning to mitigate potential disruptions and ensure continuity in operations (Handfield & Bechtel, 2002). Natural disasters and global health crises, such as the COVID-19 pandemic, have underscored the vulnerability of supply chains to unforeseen disruptions. The pandemic, in particular, exposed dependencies on single-source suppliers and highlighted the need for diversification and resilience in supply chain strategies (Ivanov & Dolgui, 2020a, 2020b). The sudden onset of the pandemic disrupted global supply chains, revealing the fragility of systems overly reliant on limited suppliers and just-in-time inventory models. This scenario prompted organizations worldwide to reconsider their supply chain strategies, incorporating greater flexibility and redundancy to withstand similar future shocks.

3.2 REVIEW OF LITERATURE

In response to these challenges, organizations are increasingly adopting proactive measures to mitigate supply chain vulnerabilities in knowledge systems. These strategies encompass a holistic approach that integrates risk assessment, technological innovation, collaboration with stakeholders, and robust contingency planning. Risk assessment frameworks aid in identifying vulnerabilities within supply chains, enabling organizations to prioritize areas for intervention and investment. By systematically evaluating the potential risks associated with each component of the supply chain, businesses can develop targeted strategies to bolster their defenses against various threats. Technological innovations are playing a pivotal role in fortifying supply chains. For instance, blockchain technology is being explored to enhance transparency and traceability across supply networks, thereby fortifying defenses against counterfeit products and unauthorized access to IP (Ivanov, Dolgui, & Sokolov, 2019). Blockchain's immutable ledger ensures that every transaction and movement of goods is recorded and verified, reducing the risk of tampering and enhancing trust among supply chain participants. Similarly, the implementation of advanced cybersecurity measures, including encryption, multi-factor authentication, and continuous monitoring, is crucial to protecting sensitive information from cyber threats.

Collaboration with suppliers and partners is essential in building resilient supply chains. Establishing clear communication channels, sharing best practices, and conducting regular audits can strengthen relationships and foster a collective approach to risk management. By engaging with suppliers on risk mitigation strategies and jointly developing contingency plans, organizations can create a more unified and robust supply chain network. Furthermore, organizations are investing in agile supply chain strategies that emphasize flexibility and responsiveness. This entails diversifying supplier bases, maintaining a surplus inventory of critical components, and leveraging digital technologies for real-time monitoring and predictive analytics (Chopra & Meindl, 2007). Digital twins, for example, allow companies to create virtual replicas of their supply chains, enabling them to simulate various scenarios and identify potential vulnerabilities before they manifest in the physical world. The complexities and implications of supply chain vulnerabilities in knowledge systems necessitate a multifaceted approach to risk management. By understanding and addressing the diverse challenges posed by cybersecurity threats, IP risks, geopolitical uncertainties, and global disruptions, organizations can safeguard their knowledge assets and sustain competitive advantage in a dynamic and interconnected world. Strategic foresight involves not only anticipating potential risks but also continuously adapting to the evolving global landscape. This requires organizations to stay informed about emerging threats, regulatory changes, and technological advancements that could impact their supply chains.

Technological innovation is another critical component of this multifaceted approach. Beyond blockchain and digital twins, artificial intelligence (AI) and machine learning (ML) are being harnessed to enhance supply chain resilience. AI and ML algorithms can analyze vast amounts of data to identify patterns and predict potential disruptions, allowing organizations to take preemptive actions. For instance, predictive maintenance powered by AI can forecast equipment failures, enabling timely repairs and reducing downtime. Collaborative efforts extend beyond immediate supply chain partners. Industry-wide collaborations and public-private partnerships can also contribute to building more resilient supply chains. Sharing information about best practices, threats, and mitigation strategies across industries and with governmental agencies can create a more comprehensive defense against systemic risks. Initiatives such as information-sharing networks and industry consortia can facilitate this level of collaboration. Thus, the contemporary global landscape demands that organizations prioritize the robustness and resilience of their knowledge systems to maintain competitive advantage. The intricacies of modern supply chains, combined with the multitude of vulnerabilities they face, require a comprehensive and proactive approach to risk management. By integrating risk assessment frameworks, embracing technological innovations, fostering collaboration with stakeholders, and developing agile supply chain strategies, organizations can navigate the complexities of the knowledge economy. This holistic approach not only mitigates risks but also leverages the full potential of knowledge systems to drive innovation, enhance productivity, and achieve sustainable growth in the global marketplace. Through strategic foresight, continuous adaptation, and a commitment to resilience, enterprises can thrive amidst the challenges and opportunities of an interconnected world.

Toyota has long been recognized for its ability to foster a high-performance knowledge-sharing network among its suppliers, which has significantly contributed to its supply chain resilience. By creating a system where suppliers could continuously learn from one another, Toyota effectively mitigates risks and disruptions, ensuring seamless operations even in times of crisis. Similarly, Procter & Gamble (P&G) has utilized data-driven supply chain strategies to enhance its manufacturing capability and customer satisfaction. P&G's knowledge management practices focus on using real-time data to monitor supply chain performance, predict disruptions, and adjust operations accordingly. This proactive approach to managing knowledge and data has enabled P&G to maintain a competitive edge and ensure a high level of resilience in their global supply chain operations. These examples highlight how effective knowledge management can help companies navigate supply chain vulnerabilities by fostering collaboration and data utilization.

3.3 UNDERSTANDING KNOWLEDGE SYSTEMS IN SUPPLY CHAINS

Knowledge systems within supply chains encompass the complex mechanisms through which organizations acquire, generate, manage, and disseminate knowledge. They are indispensable in industries such as technology, pharmaceuticals, automotive, and aerospace, where innovation and IP are fundamental. According to various scholars, including Peter Drucker (1993) and Paul D. Cousins and Handfield (1996), these systems facilitate collaboration among internal teams, external partners, suppliers, and sometimes competitors. Such collaborations are crucial for fostering continuous learning, driving innovation, and creating value across the supply chain. In the context of modern industry, effective knowledge management ensures that organizations can leverage their intellectual assets optimally. This capability is essential for maintaining competitiveness and responding adeptly to market dynamics and technological advancements. As Drucker (1993) emphasized, we are living in a "knowledge economy" where the primary resource is not capital, labor, or land, but knowledge. Consequently, the management of knowledge resources is pivotal for organizational success.

Within supply chains, knowledge systems play a pivotal role in several key areas:

1. **Facilitating Innovation**
   In industries like technology and pharmaceuticals, the rapid pace of innovation is a competitive necessity. Knowledge systems enable organizations to share research and development insights quickly and efficiently, fostering a collaborative environment that accelerates innovation (Nonaka & Takeuchi, 1995; Davenport & Prusak, 1998). For example, cross-functional teams can leverage centralized databases to access the latest scientific findings or technological breakthroughs, thereby reducing the time to market for new products (Leonard-Barton, 1995; Cohen & Levinthal, 1990). These systems facilitate the rapid dissemination of critical information, enabling teams to build on existing knowledge and make informed decisions that drive innovation forward.
2. **Optimizing Production and Distribution**
   Knowledge systems streamline production processes by integrating information from various stages of the supply chain (Gunasekaran & Ngai, 2004; Li et al., 2006). This integration allows for real-time monitoring and adjustments, enhancing efficiency and reducing waste. For instance, in the automotive industry, knowledge systems help synchronize the supply chain by sharing production schedules and inventory levels with suppliers, ensuring that parts and components are delivered just in time, thus minimizing inventory costs (Holweg, 2005; Womack, Jones, & Roos, 1990).
3. **Enhancing Intellectual Property Management**
   IP is a critical asset in knowledge-intensive industries. Effective knowledge systems safeguard IP by implementing robust data security measures and access controls (Chesbrough, 2003; Jaffe & Lerner, 2004). Additionally, these systems facilitate the management of IP portfolios, enabling organiza-

tions to track patents, trademarks, and copyrights effectively (Hall, 2005; Rivette & Kline, 2000). By ensuring that IP is protected and properly managed, companies can prevent unauthorized use and capitalize on their innovations (Teece, 2000).

4. **Enabling Strategic Decision-Making**
   Knowledge systems provide executives and managers with the information needed to make informed strategic decisions (Davenport & Prusak, 1998; Alavi & Leidner, 2001). By consolidating data from various sources, these systems offer a comprehensive view of the supply chain, highlighting potential risks and opportunities (Gunasekaran et al., 2004). For example, predictive analytics can forecast demand trends, allowing organizations to adjust their production and inventory strategies proactively (Chae, 2015; Waller & Fawcett, 2013).
5. **Supporting Collaboration and Learning**
   Collaboration is essential for continuous improvement and innovation. Knowledge systems facilitate communication and information sharing among supply chain partners, including suppliers, distributors, and customers (Lee, 2000; Simatupang & Sridharan, 2002). This collaboration enhances collective learning, driving improvements across the supply chain. For example, in the aerospace industry, manufacturers often collaborate with suppliers to develop new materials or components, sharing knowledge to achieve common goals (Doz & Hamel, 1998; Takeishi, 2001).

Despite their benefits, managing knowledge systems within supply chains presents several challenges. One major issue is ensuring data accuracy and integrity. Inaccurate or outdated information can lead to poor decision-making and inefficiencies (Redman, 1998; Wang & Strong, 1996). To address this, organizations must implement rigorous data governance practices, ensuring that data is collected, validated, and updated regularly (Khatri & Brown, 2010). Another challenge is cybersecurity. As knowledge systems become increasingly interconnected, they become more vulnerable to cyber threats. Protecting sensitive information from cyberattacks is crucial (Gordon & Loeb, 2002a, 2002b; Von Solms & Van Niekerk, 2013). This involves deploying advanced security measures such as encryption, multi-factor authentication, and continuous monitoring to safeguard data (Schneier, 2015).

Additionally, fostering a culture of knowledge sharing can be difficult, particularly in competitive environments. Organizations must encourage collaboration by creating incentives for knowledge sharing and ensuring that employees understand the value of collective intelligence (Davenport & Prusak, 1998; Alavi & Leidner, 2001). This can be achieved through training programs, performance metrics, and recognition schemes that reward knowledge contributions (Bartol & Srivastava, 2002). Moreover, the global nature of supply chains introduces complexities related to differing legal frameworks and cultural differences. For instance, IP laws vary significantly across countries, affecting how organizations manage and protect their intellectual assets internationally (Maskus, 2000; Yang & Clarke, 2005). Companies must navigate these differences carefully, ensuring compliance with local regulations while protecting their interests (Rivette & Kline, 2000).

Thus, knowledge systems within supply chains are indispensable for fostering innovation, enhancing productivity, and maintaining competitive advantage. By facilitating collaboration, optimizing processes, and protecting intellectual assets, these systems enable organizations to thrive in the knowledge economy (Nonaka & Takeuchi, 1995; Drucker, 1993). However, managing these systems effectively requires addressing challenges related to data accuracy, cybersecurity, and cultural barriers. Through strategic investment in technology and fostering a collaborative culture, organizations can harness the full potential of their knowledge systems, driving sustainable growth and success in an increasingly interconnected world (Chesbrough, 2003; Teece, 2000).

3.4 TYPES OF VULNERABILITIES

3.4.1 Cybersecurity Threats

In contemporary business environments, knowledge systems heavily rely on digital platforms, cloud computing, and interconnected networks to enhance collaboration and streamline data exchange. However, these technological dependencies also introduce significant vulnerabilities, exposing organizations to various cyber threats such as data breaches, ransomware attacks, and IP theft, which can severely impact operations across industries. The implications of such cyber threats are profound. A successful cyberattack can compromise sensitive information, disrupt critical operations, erode stakeholder trust, and cause substantial financial and reputational damage. For instance, in the healthcare sector, the digitization of medical records has heightened cybersecurity concerns due to the sensitivity of patient data. Authors like Peter Drucker (1993) emphasize the importance of managing knowledge assets securely in the digital age, highlighting the need for robust cybersecurity measures and proactive risk management strategies to safeguard organizational integrity and maintain competitive advantage (Rosenzweig, 2019a, 2019b).

Effective knowledge systems are crucial for organizations to leverage their intellectual assets optimally, maintaining competitiveness and responding adeptly to market dynamics and technological advancements. Collaboration is essential for continuous improvement and innovation. Knowledge systems facilitate communication and information sharing among supply chain partners, including suppliers, distributors, and customers (Lee, 2000; Simatupang & Sridharan, 2002). This collaboration enhances collective learning, driving improvements across the supply chain. For example, in the aerospace industry, manufacturers often collaborate with suppliers to develop new materials or components, sharing knowledge to achieve common goals (Doz & Hamel, 1998; Takeishi, 2001). Despite their benefits, managing knowledge systems within supply chains presents several challenges. One major issue is ensuring data accuracy and integrity. Inaccurate or outdated information can lead to poor decision-making and inefficiencies (Redman, 1998; Wang & Strong, 1996). To address this, organizations must implement rigorous data governance practices, ensuring that data is collected, validated, and updated regularly (Khatri & Brown, 2010). Another challenge is cybersecurity. As knowledge systems become increasingly interconnected, they become more vulnerable to cyber threats. Protecting sensitive information from cyberattacks is crucial (Gordon & Loeb, 2002a, 2002b; Von Solms & Van Niekerk, 2013). This involves deploying advanced security measures such as encryption, multi-factor authentication, and continuous monitoring to safeguard data (Schneier, 2015).

Additionally, fostering a culture of knowledge sharing can be difficult, particularly in competitive environments. Organizations must encourage collaboration by creating incentives for knowledge sharing and ensuring that employees understand the value of collective intelligence (Davenport & Prusak, 1998; Alavi & Leidner, 2001). This can be achieved through training programs, performance metrics, and recognition schemes that reward knowledge contributions (Bartol & Srivastava, 2002). Moreover, the global nature of supply chains introduces complexities related to differing legal frameworks and cultural differences. For instance, IP laws vary significantly across countries, affecting how organizations manage and protect their intellectual assets internationally (Maskus, 2000; Yang & Clarke, 2005). Companies must navigate these differences carefully, ensuring compliance with local regulations while protecting their interests (Rivette & Kline, 2000). Thus, knowledge systems within supply chains are indispensable for fostering innovation, enhancing productivity, and maintaining competitive advantage. By facilitating collaboration, optimizing processes, and protecting intellectual assets, these systems enable organizations to thrive in the knowledge economy (Nonaka & Takeuchi, 1995; Drucker, 1993). However, managing these systems effectively requires addressing challenges related to data accuracy, cybersecurity, and cultural barriers. Through strategic investment in technology and fostering a collaborative culture, organizations can harness the full potential of their knowledge systems, driving sustainable growth and success in an increasingly interconnected world (Chesbrough, 2003; Teece, 2000).

3.4.2 Dependency on External Suppliers

In knowledge-intensive industries, collaboration with specialized suppliers and partners is essential for accessing critical components, advanced technologies, and specialized expertise. However, disruptions within the supply chain present significant challenges. Geopolitical tensions, natural disasters, and economic instability can all interrupt the flow of knowledge and innovation processes, impacting organizational resilience and competitive advantage. Supply chain disruptions can lead to substantial setbacks such as delays in product development, increased costs, and reduced competitiveness. Additionally, over-reliance on a limited number of suppliers increases vulnerability to supplier failures or abrupt shifts in market dynamics, as noted by scholars like Christopher and Lee. A poignant example is the semiconductor industry, which encountered severe supply chain disruptions during the COVID-19 pandemic. These disruptions hindered the availability of critical components necessary for various technologies, affecting global production and highlighting vulnerabilities in supply chain dependencies (Ivanov & Dolgui, 2020a, 2020b). The pandemic exposed the fragility of global supply chains and underscored the importance of having robust strategies to mitigate such risks.

To address these challenges, organizations must adopt comprehensive supply chain management strategies that emphasize diversification of suppliers, proactive risk assessment, and agile response mechanisms. Diversification involves sourcing critical components from multiple suppliers across different regions to reduce dependency on any single source. This strategy not only mitigates risks associated with geopolitical tensions and natural disasters but also ensures a steady flow of essential materials and components. Proactive risk assessment is another crucial aspect of effective supply chain management. Organizations need to identify potential vulnerabilities within their supply chains and develop contingency plans to address them. This includes regular audits of suppliers, evaluating geopolitical risks, and monitoring economic indicators that could affect supply chain stability. By understanding and preparing for potential disruptions, companies can reduce the impact of unforeseen events on their operations.

Agile response mechanisms are essential for maintaining continuity in knowledge flows and sustaining innovation in dynamic and unpredictable global environments. This involves implementing flexible production processes and maintaining strategic stockpiles of critical components to buffer against supply chain interruptions. Agile supply chain management also leverages advanced technologies such as AI and predictive analytics to anticipate disruptions and optimize responses in real time. Furthermore, collaboration with suppliers and partners can be enhanced through effective communication and information sharing. Establishing strong relationships with key suppliers and fostering a culture of transparency can help organizations respond more effectively to disruptions. Joint risk management initiatives and collaborative planning can also improve supply chain resilience by ensuring all parties are aligned in their strategies and objectives.

Investing in digital supply chain technologies is another critical measure. Digital platforms enable real-time tracking of supply chain activities, providing visibility into potential issues before they escalate into major disruptions. Technologies like blockchain can enhance transparency and traceability, ensuring the integrity of information and reducing the risk of fraud or errors. Thus, while supply chain disruptions pose significant challenges to knowledge-intensive industries, adopting robust supply chain management strategies can mitigate these risks. Diversifying suppliers, conducting proactive risk assessments, implementing agile response mechanisms, and leveraging digital technologies are essential steps for maintaining continuity in knowledge flows and sustaining innovation. By addressing these challenges, organizations can enhance their resilience and competitive advantage in an increasingly dynamic and unpredictable global environment.

3.4.3 Technological Advancements and Obsolescence

In today's rapidly evolving technological landscape, sustained competitiveness hinges on organizations' ability to continually invest in updating infrastructure, tools, and workforce capabilities. Management theorists like Peter Drucker (1993) argue that failure to swiftly adopt emerging technologies can render knowledge systems obsolete, diminishing their effectiveness over time. This poses a significant risk, as competitors who adeptly harness new technologies gain a strategic edge. Technological obsolescence not only impedes innovation but also constrains growth opportunities and diminishes market relevance. This challenge is particularly pronounced in traditional manufacturing sectors transitioning to Industry 4.0 technologies. Ivanov and Dolgui (2020a, 2020b) highlight the complexities involved in integrating automation, the Internet of Things (IoT), and AI due to legacy systems and workforce skill gaps. For instance, adopting advanced manufacturing technologies often requires substantial investments in retraining the workforce and upgrading existing infrastructure.

Organizations must navigate these challenges to effectively leverage technological advancements, optimize operational efficiency, and sustain competitiveness in a rapidly transforming global marketplace. This involves a multifaceted approach, starting with the alignment of strategic objectives with technological investments. Companies need to conduct thorough assessments of their existing systems and identify areas where new technologies can create the most significant impact. Investing in workforce development is crucial. As technologies evolve, the skills required to manage and operate new systems change. Organizations must prioritize continuous learning and professional development to ensure their employees are equipped with the necessary skills. This can be achieved through targeted training programs, partnerships with educational institutions, and fostering a culture of innovation and adaptability within the organization.

Upgrading infrastructure is another critical component. Legacy systems often lack the compatibility and flexibility required to integrate with modern technologies. Therefore, organizations need to invest in new infrastructure that supports scalability and interoperability. This might include transitioning to cloud-based platforms, adopting modular systems that allow for incremental upgrades, and ensuring robust cybersecurity measures to protect against emerging threats. The integration of new technologies should also be approached with a focus on collaboration and communication. Cross-functional teams can play a pivotal role in facilitating smooth transitions by bringing together diverse perspectives and expertise. This collaborative approach helps in identifying potential challenges early and devising effective solutions. For instance, involving IT specialists, operations managers, and frontline workers in the planning and implementation phases can ensure that technological upgrades align with practical operational needs.

Furthermore, organizations should adopt a proactive stance toward technological trends and disruptions. Staying informed about emerging technologies and industry developments can provide a competitive advantage. This includes engaging with industry forums, participating in pilot projects, and fostering relationships with technology providers and startups. By staying ahead of the curve, organizations can anticipate changes and adapt more swiftly. To summarize, in a fast-evolving technological landscape, organizations must continually invest in updating infrastructure, tools, and workforce capabilities to maintain competitiveness. Addressing the challenges of technological obsolescence requires a strategic approach that includes aligning objectives with technological investments, prioritizing workforce development, upgrading infrastructure, fostering collaboration, and staying informed about industry trends. Through these efforts, organizations can leverage technological advancements to drive innovation, optimize operations, and sustain their competitive edge in a dynamic global marketplace.

3.4.4 Regulatory and Compliance Requirements

Navigating the regulatory landscape presents a formidable challenge for knowledge systems, which must contend with diverse requirements such as data privacy laws, IP rights, and industry standards that vary by region and jurisdiction. Scholars like Maskus (2000) emphasize the critical importance of compliance to avoid legal penalties, fines, lawsuits, and reputational harm. Operating across multiple markets amplifies these challenges, especially when faced with conflicting or evolving regulatory frameworks. For instance, the General Data Protection Regulation (GDPR) implemented in Europe imposes rigorous standards for data protection, significantly impacting how organizations handle and exchange customer and employee data worldwide (Rosenzweig, 2019a, 2019b). The complexities of compliance necessitate robust strategies that integrate legal expertise, technological solutions for data management, and ongoing regulatory monitoring. Organizations must adopt a comprehensive approach to ensure compliance across various jurisdictions and to navigate the intricacies of international regulations effectively.

Legal expertise is essential for interpreting and applying regulatory requirements accurately. Companies should invest in legal teams or consult with external legal advisors who specialize in the relevant laws and regulations of each market they operate in. These legal experts can provide guidance on best practices, help in drafting compliant policies, and represent the organization in legal matters. Technological solutions for data management play a crucial role in achieving and maintaining compliance. Advanced data management systems can automate the enforcement of compliance policies, monitor data access and usage, and provide audit trails. For example, implementing data encryption, anonymization, and secure access controls can help protect sensitive information and comply with stringent data privacy laws like GDPR. Additionally, deploying compliance management software can streamline the process of tracking and adhering to various regulatory requirements across different regions.

Ongoing regulatory monitoring is vital for staying current with evolving laws and standards. Organizations must establish mechanisms for continuous surveillance of regulatory changes and updates. This could involve subscribing to legal and regulatory information services, participating in industry associations, and maintaining active communication with regulatory bodies. By staying informed about new and amended regulations, companies can swiftly adapt their practices to remain compliant. Proactively addressing compliance challenges offers several benefits. First, it mitigates the risk of legal penalties and financial losses associated with non-compliance. Second, it enhances operational transparency, which can improve relationships with regulators and stakeholders. Third, maintaining compliance fosters trust among customers and partners, as it demonstrates a commitment to ethical practices and the protection of personal and proprietary information.

Furthermore, integrating compliance into the organizational culture is crucial. Employees at all levels should be educated about the importance of regulatory adherence and trained on the specific requirements relevant to their roles. Regular training sessions, clear communication of compliance policies, and establishing a whistleblower mechanism can help reinforce a culture of compliance. Thus, navigating the regulatory landscape is a complex but essential task for knowledge systems, requiring a blend of legal expertise, advanced technological solutions, and vigilant regulatory monitoring. By adopting a proactive and comprehensive approach to compliance, organizations can mitigate risks, enhance transparency, and maintain stakeholder trust in an increasingly regulated global environment. This strategic focus not only safeguards the organization from legal repercussions but also ensures sustainable operations and competitive advantage in the global marketplace.

3.5 STRATEGIES TO MITIGATE SUPPLY CHAIN VULNERABILITIES

3.5.1 Enhancing Cybersecurity Measures

In today's interconnected digital landscape, implementing rigorous cybersecurity measures is essential to protect knowledge systems from malicious activities. Scholars such as Rosenzweig (2019a, b) emphasize the critical role of encryption, multi-factor authentication, and frequent vulnerability assessments in bolstering organizational defenses against cyber threats. These protocols safeguard sensitive data and IP, mitigating the risk of breaches that could compromise organizational integrity and operational continuity. Equally important is the education of employees and stakeholders on cybersecurity best practices. Ivanov and Dolgui (2020a, 2020b) highlight the significance of comprehensive training initiatives and awareness campaigns aimed at cultivating a security-conscious culture within the organization. By fostering an understanding of potential threats and promoting proactive responses, these efforts empower individuals to contribute actively to cyber resilience. Through a holistic approach that combines robust technological solutions with ongoing education and awareness-building, organizations can significantly enhance their cybersecurity posture. This integrated strategy not only strengthens defenses against cyberattacks but also instills confidence among stakeholders and reinforces the organization's commitment to safeguarding valuable assets in an increasingly complex threat landscape. By prioritizing both technological and human elements of cybersecurity, organizations can better navigate the challenges of the digital age and protect their critical knowledge systems.

Brusset and Teller (2022) argue that implementing multi-layered cybersecurity measures is paramount for protecting knowledge systems from cyber threats. Encryption, multi-factor authentication, and regular security audits are essential tools in maintaining secure operations. Furthermore, Rosenzweig (2019a, b) highlights that cyberattacks on supply chains can significantly disrupt operations, leading to knowledge loss and reduced business continuity. In 2017, Maersk, the global shipping giant, experienced a significant cybersecurity breach due to the NotPetya ransomware attack, which crippled their operations worldwide. In response, Maersk implemented a comprehensive cybersecurity overhaul that included advanced encryption methods, multi-factor authentication, and regular vulnerability assessments. They also instituted extensive training programs for employees to ensure a security-aware culture across the organization. As a result, Maersk improved its cybersecurity posture and was able to recover swiftly, demonstrating how robust cybersecurity measures can protect critical knowledge systems and ensure operational continuity.

3.5.2 Diversifying Supply Chains

Reducing dependency on single suppliers is crucial for mitigating risks in knowledge systems. Scholars like Christopher and Lee emphasize the importance of identifying and qualifying alternative suppliers and partners. Diversifying sources helps organizations maintain continuity in the face of supplier failures or geopolitical disruptions, thereby safeguarding supply chain resilience and operational stability. Establishing comprehensive contingency plans and risk management strategies is equally essential. According to Handfield and Bechtel (2002), these strategies involve proactive measures such as mapping supply chain vulnerabilities, developing backup supplier relationships, and implementing agile response protocols. By anticipating potential disruptions and preparing effective response mechanisms, organizations can minimize the impact of unforeseen events on knowledge flows and innovation processes.

Ivanov et al. (2021) emphasize that diversifying supply chains has become increasingly critical to ensure resilience in the face of global disruptions, such as the COVID-19 pandemic and geopolitical conflicts. Diversification helps reduce reliance on single-source suppliers, which can make organizations vulnerable to sudden interruptions in knowledge and material flows. Moreover, Azadegan et al. (2020) found that supply chains diversified across multiple regions are more adaptive and exhibit faster recovery post-disruption. Sharma et al. (2022) suggest using AI-driven supplier analytics and monitoring tools to continuously evaluate supplier performance and risks. This tech-enabled approach provides real-time visibility, allowing businesses to quickly pivot to alternative suppliers when required. Ongoing monitoring and periodic reassessment of supplier networks are also recommended. This ensures that contingency plans remain relevant and responsive to evolving market dynamics and geopolitical shifts. Regular evaluations help organizations identify emerging risks and adjust their strategies, enhancing their capacity to navigate an increasingly complex global landscape. Additionally, fostering strong relationships with multiple suppliers can provide a buffer against disruptions. Collaborative partnerships encourage shared responsibility and transparency, facilitating better coordination and quicker responses during crises. This cooperative approach not only strengthens the supply chain but also promotes continuous improvement and innovation through the exchange of ideas and best practices.

Investing in advanced technologies such as blockchain and IoT can further enhance supply chain visibility and traceability. These technologies enable real-time monitoring of supply chain activities, allowing for prompt identification of potential issues and swift corrective actions. By leveraging digital tools, organizations can improve their agility and resilience, ensuring that knowledge systems remain robust and effective even in the face of disruptions. Ultimately, integrating these strategies into supply chain management practices enhances an organization's adaptive capacity and helps maintain competitive advantage in a dynamic global environment characterized by uncertainty and volatility. By diversifying suppliers, developing comprehensive contingency plans, fostering collaborative relationships, and utilizing advanced technologies, organizations can build resilient knowledge systems capable of sustaining innovation and operational continuity despite external challenges.

Unilever, a leading consumer goods company, faced supply chain disruptions during the COVID-19 pandemic due to its reliance on single-source suppliers. To mitigate this risk, Unilever diversified its supplier base by identifying alternative suppliers across different geographic regions. They also established strategic partnerships with local suppliers, ensuring continuity in production despite lockdowns. This diversification allowed Unilever to maintain product availability and respond quickly to changing market demands, illustrating the importance of a flexible supplier network in safeguarding knowledge flows and innovation.

3.5.3 Adopting Agile and Flexible Technologies

In today's business environment, organizations increasingly recognize the imperative of investing in adaptable technologies to enhance their supply chain capabilities. Scholars such as Chopra and Meindl (2007) underscore the importance of scalable and interoperable technologies that enable swift deployment and adaptation to shift market conditions. These technologies not only boost operational efficiency but also bolster overall agility and responsiveness. Cloud computing, IoT, blockchain, and AI are pivotal in this regard. Ivanov, Dolgui, and Sokolov (2019) advocate for their adoption to improve visibility, traceability, and decision-making throughout supply chains. Cloud platforms offer scalable infrastructure for efficient data storage and processing, while IoT facilitates real-time monitoring of assets and processes. Blockchain technology ensures transaction transparency and security, mitigating risks related to fraud and data manipulation. AI algorithms analyze extensive datasets to extract actionable insights, optimizing resource allocation and enhancing demand forecasting accuracy.

Ivanov, Dolgui, Sokolov, and Ivanova (2021) advocate for investing in IoT and blockchain to enhance the traceability of goods and reduce disruptions in knowledge systems. Similarly, Qin et al. (2022) demonstrate that AI and ML models can be employed to forecast demand fluctuations and optimize resource allocation in real time, creating a more responsive and agile supply chain. Chopra and Meindl (2020) highlight that companies embracing cloud computing and AI have seen substantial improvements in decision-making speed, allowing for quicker adjustments to unexpected market changes. This adaptability directly supports resilience in knowledge-driven industries, where rapid access to data is essential. By embracing these technologies, organizations can fortify their competitive position by fostering innovation, bolstering operational resilience, and meeting evolving customer expectations in a rapidly evolving global marketplace marked by digital transformation. These technological investments not only streamline supply chain operations but also empower organizations to respond promptly to market fluctuations and customer demands, thereby sustaining growth and maintaining leadership in their respective industries. Zara, the fashion retailer, has long been known for its agile supply chain practices. By adopting IoT and data analytics, Zara continuously monitors customer preferences and inventory levels in real time. This data-driven approach allows Zara to quickly adapt its production and distribution strategies, ensuring that the latest trends are reflected in-store within weeks. Their implementation of advanced technologies not only streamlines operations but also enhances customer satisfaction and drives innovation in their supply chain management practices.

3.6 STRENGTHENING REGULATORY COMPLIANCE

In navigating the complexities of regulatory environments and industry standards relevant to knowledge systems, organizations must cultivate a thorough understanding to mitigate risks and ensure operational compliance. Authors such as Maskus (2000) emphasize the importance of developing a comprehensive grasp of these requirements to safeguard IP and maintain legal integrity across global operations. Establishing robust governance frameworks and compliance programs is pivotal in this regard. These frameworks, as advocated by Rosenzweig (2019a, b), serve to institutionalize processes that monitor, assess, and enforce adherence to legal obligations and ethical standards. This includes protocols for data privacy, IP protection, and cybersecurity measures, ensuring alignment with evolving regulatory landscapes.

Linton and Vakil (2022) stress that organizations must adopt comprehensive compliance strategies to avoid legal risks and maintain operational continuity. Recent studies show that non-compliance with international data privacy laws, such as the GDPR, can result in substantial fines and reputational damage. Kerry and Morris (2021) highlight how supply chains in the pharmaceutical industry have responded to regulatory challenges by establishing centralized compliance management systems, ensuring that every node in the supply chain aligns with evolving laws.

Additionally, Maskus (2021) emphasizes the importance of IP protection in maintaining the integrity of knowledge systems. As supply chains become more digitized, establishing clear governance protocols around IP is key to protecting valuable organizational knowledge. By integrating these governance structures, organizations not only mitigate legal and reputational risks but also cultivate a culture of ethical conduct and transparency. This approach enhances stakeholder trust and reinforces organizational resilience in the face of regulatory scrutiny and industry scrutiny. It also supports strategic decision-making by providing a framework for evaluating and responding to regulatory changes, thereby sustaining competitive advantage in global markets characterized by stringent compliance requirements.

Pfizer, a global pharmaceutical company, operates in a highly regulated environment requiring strict compliance with various health and safety standards. To manage these complexities, Pfizer developed a comprehensive compliance management system that integrates real-time monitoring of regulatory changes. By establishing robust governance frameworks and training programs, Pfizer ensures that all employees are aware of compliance requirements, which minimizes legal risks and enhances operational integrity. This proactive approach to regulatory compliance has allowed Pfizer to maintain its reputation and operate effectively across multiple jurisdictions.

3.7 PROMOTING COLLABORATION AND KNOWLEDGE SHARING

In today's interconnected business environment, fostering collaborative partnerships across industry peers, academia, research institutions, and governmental agencies is instrumental in addressing supply chain vulnerabilities and enhancing industry resilience. Authors such as Handfield and Bechtel (2002) emphasize the strategic advantages of forming alliances to pool resources, share expertise, and collectively innovate in supply chain management. Collaborative partnerships enable organizations to share best practices, lessons learned, and emerging technologies. By leveraging collective insights and experiences, stakeholders can proactively identify and mitigate supply chain risks. This collaborative approach is crucial in navigating challenges such as geopolitical tensions, natural disasters, and cybersecurity threats that disrupt global supply chains.

Kleindorfer and Saad (2020) emphasize the role of knowledge-sharing platforms in improving visibility across the supply chain. These platforms enable organizations to collaborate more effectively, ensuring that all participants remain informed of potential risks and can adapt proactively. Ivanov et al. (2021) argue that cross-industry collaboration and open knowledge sharing have become vital in mitigating supply chain vulnerabilities. Recent studies by Manning et al. (2022) demonstrate that strategic partnerships between organizations, academic institutions, and government bodies foster a more innovative and resilient supply chain ecosystem. Collaborative networks can help organizations respond swiftly to emerging threats and share best practices. Furthermore, partnerships facilitate knowledge exchange and capacity building, fostering a culture of continuous learning and improvement. Ivanov, Dolgui, and Sokolov (2019) advocate for collaborative efforts to enhance supply chain resilience through digital technologies like blockchain and AI. By aligning interests and pooling resources, organizations can achieve greater efficiency, agility, and sustainability in their supply chain operations, ultimately strengthening their competitive position in the global marketplace.

In the aerospace sector, companies like Boeing have recognized the value of collaborative partnerships to enhance supply chain resilience. Boeing has established collaborative networks with suppliers, academic institutions, and research organizations to share best practices and innovative technologies. For instance, through the Advanced Manufacturing Initiative, Boeing collaborates with suppliers to develop advanced manufacturing techniques and materials. This collaborative effort not only improves supply chain transparency but also drives innovation and strengthens the overall resilience of the aerospace supply chain.

3.8 CONCLUSION

This chapter emphasizes that supply chain vulnerabilities significantly impact the resilience and effectiveness of knowledge systems. Addressing these risks requires proactive strategies, including robust cybersecurity measures, supply chain diversification, agile technology adoption, regulatory compliance, and collaboration. Implementing strong cybersecurity protocols, such as encryption and multi-factor authentication, protects sensitive data and IP. Diversifying supply sources reduces dependency and increases resilience against disruptions, while agile technologies like cloud computing, IoT, and AI enhance operational visibility and responsiveness. Regulatory compliance, combined with cross-industry collaboration, fosters transparency and trust, while internal knowledge-sharing cultures strengthen organizational resilience. By integrating these strategies, organizations can safeguard knowledge assets, promote innovation, and maintain a competitive edge in a dynamic marketplace.

REFERENCES

Alavi, M., and D. E. Leidner. "Knowledge Management and Knowledge Management Systems: Conceptual Foundations and Research Issues." *MIS Quarterly* 25, no. 1 (2001): 107–136.

Azadegan, A., P. Patel, G. Zang, and K. Linderman. "Resilient Supply Chain Management: Strategies and Performance Metrics." *Journal of Supply Chain Management* 56, no. 1 (2020): 38–50.

Bartol, K. M., and A. Srivastava. "Encouraging Knowledge Sharing: The Role of Organizational Reward Systems." *Journal of Leadership & Organizational Studies* 9, no. 1 (2002): 64–76.

Brusset, X., and C. Teller. "Digital Supply Chain Security: A Cybersecurity Framework for Preventing Disruptions." *Supply Chain Management Review* 48, no. 4 (2022): 32–44.

Chae, B. "Insights from Hashtag #SupplyChain and Twitter Analytics: Considering Twitter and Twitter Data for Supply Chain Practice and Research." *International Journal of Production Economics* 165 (2015): 247–259.

Chesbrough, H. W. *Open Innovation: The New Imperative for Creating and Profiting from Technology*. Harvard Business School Press, 2003.

Chopra, S., and P. Meindl. *Supply Chain Management: Strategy, Planning, and Operation*. Pearson Prentice Hall, 2007.

Chopra, S., and P. Meindl. *Supply Chain Management: Strategy, Planning, and Operation*. Pearson, 2020.

Cohen, W. M., and D. A. Levinthal. "Absorptive Capacity: A New Perspective on Learning and Innovation." *Administrative Science Quarterly* 35, no. 1 (1990): 128–152.

Cousins, P. D., and R. B. Handfield. "Supply Chain Management: A Strategic Perspective." *International Journal of Physical Distribution & Logistics Management* 26, no. 8 (1996): 15–28.

Davenport, T. H., and L. Prusak. *Working Knowledge: How Organizations Manage What They Know*. Harvard Business School Press, 1998.

Doz, Y. L., and G. Hamel. *Alliance Advantage: The Art of Creating Value Through Partnering*. Harvard Business School Press, 1998.

Drucker, P. *Post-Capitalist Society*. HarperBusiness, 1993.

Gordon, L. A., and M. P. Loeb. "The Economics of Information Security Investment." *ACM Transactions on Information and System Security* 5, no. 4 (2002a): 438–457.

Gordon, L. A., and M. P. Loeb. *Managing Cybersecurity Resources: A Cost-Benefit Analysis*. McGraw-Hill, 2002b.

Gunasekaran, A., and E. W. T. Ngai. "Information Systems in Supply Chain Integration and Management." *European Journal of Operational Research* 159, no. 2 (2004): 269–295.

Gunasekaran, A., C. Patel, and R. E. McGaughey. "A Framework for Supply Chain Performance Measurement." *International Journal of Production Economics* 87, no. 3 (2004): 333–347.

Hall, B. H. "Exploring the Patent Explosion." *Journal of Technology Transfer* 30, nos. 1–2 (2005): 35–48.

Handfield, R. B., and C. Bechtel. "The Role of Trust and Relationship Structure in Improving Supply Chain Responsiveness." *Industrial Marketing Management* 31, no. 4 (2002): 367–382.

Holweg, M. "The Three Dimensions of Responsiveness." *International Journal of Operations & Production Management* 25, no. 7 (2005): 603–622.

Ivanov, D., and A. Dolgui. "A Digital Supply Chain Twin for Managing the Disruption Risks and Resilience in the Era of Industry 4.0." *Production Planning & Control* 31, no. 11–12 (2020a): 938–953.

Ivanov, D., and A. Dolgui. "Viability of Intertwined Supply Networks: Extending the Supply Chain Resilience Angles Towards Survivability." *International Journal of Production Research* 58, no. 10 (2020b): 2904–2915.

Ivanov, D., A. Dolgui, and B. Sokolov. "The Impact of Digital Technology and Industry 4.0 on the Ripple Effect and Supply Chain Risk Analytics." *International Journal of Production Research* 57, no. 3 (2019): 829–846.

Ivanov, D., A. Dolgui, B. Sokolov, and M. Ivanova. "Supply Chain Viability and the COVID-19 Pandemic: A Conceptual and Formal Generalisation of Four Major Adaptation Strategies." *International Journal of Production Research* 59, no. 12 (2021): 3535–3552.

Jaffe, A. B., and J. Lerner. *Innovation and Its Discontents: How Our Broken Patent System is Endangering Innovation and Progress, and What to Do About It*. Princeton University Press, 2004.

Kerry, C. F., and B. Morris. "Digital Supply Chains and GDPR Compliance: A Strategic Framework." *Journal of International Business Law* 40, no. 5 (2021): 113–129.

Khatri, V., and C. V. Brown. "Designing Data Governance." *Communications of the ACM* 53, no. 1 (2010): 148–152.

Kleindorfer, P. R., and G. H. Saad. "Managing Disruption Risks in Supply Chains Through Knowledge Sharing." *Journal of Production and Operations Management* 29, no. 7 (2020): 821–831.

Lee, H. L. "Creating Value Through Supply Chain Integration." *Supply Chain Management Review* 4, no. 4 (2000): 30–36.

Leonard-Barton, D. *Wellsprings of Knowledge: Building and Sustaining the Sources of Innovation*. Harvard Business School Press, 1995.

Li, S., B. Ragu-Nathan, T. S. Ragu-Nathan, and S. S. Rao. "The Impact of Supply Chain Management Practices on Competitive Advantage and Organizational Performance." *Omega* 34, no. 2 (2006): 107–124.

Linton, T., and B. Vakil. "Supply Chain Regulatory Compliance and Global Trade Risks." *Harvard Business Review* 100, no. 2 (2022): 50–65.

Manning, S., F. Boons, and J. Von Hagen. "Supply Chain Collaboration and Knowledge Sharing for Resilience." *Global Supply Chain Management Journal* 9, no. 3 (2022): 47–66.

Maskus, K. E. *Intellectual Property Rights in the Global Economy*. Peterson Institute for International Economics, 2000.

Maskus, K. E. *Intellectual Property Rights in the Global Economy*. Institute for International Economics, 2021.

Nonaka, I., and H. Takeuchi. *The Knowledge-Creating Company: How Japanese Companies Create the Dynamics of Innovation*. Oxford University Press, 1995.

Qin, J., Y. Liu, Z. Zhou, and Y. Fan. "AI-Driven Supply Chain Agility in Knowledge Systems: A Case Study Approach." *Journal of Artificial Intelligence Research* 57, no. 3 (2022): 22–34.

Redman, T. C. "The Impact of Poor Data Quality on the Typical Enterprise." *Communications of the ACM* 41, no. 2 (1998): 79–82.

Rivette, K. G., and D. Kline. *Rembrandts in the Attic: Unlocking the Hidden Value of Patents*. Harvard Business School Press, 2000.

Rosenzweig, P. *Cybersecurity and Cyberwar: What Everyone Needs to Know*. Oxford University Press, 2019a.

Rosenzweig, P. M. "The Cyber Threat to Global Supply Chains." *Information Security Journal: A Global Perspective* 28, no. 1 (2019b): 3–16.

Schneier, B. *Data and Goliath: The Hidden Battles to Collect Your Data and Control Your World*. W. W. Norton & Company, 2015.

Sharma, P., D. Ivanov, and A. Behl. "AI-Enabled Supply Chain Risk Management: A Digital Transformation Approach for Diversification." *Journal of Business Logistics* 43, no. 2 (2022): 123–142.

Simatupang, T. M., and R. Sridharan. "The Collaborative Supply Chain." *The International Journal of Logistics Management* 13, no. 1 (2002): 15–30.

Takeishi, A. "Bridging Inter- and Intra-Firm Boundaries: Management of Supplier Involvement in Automobile Product Development." *Strategic Management Journal* 22, no. 5 (2001): 403–433.

Teece, D. J. *Managing Intellectual Capital: Organizational, Strategic, and Policy Dimensions*. Oxford University Press, 2000.

Von Solms, R., and J. Van Niekerk. "From Information Security to Cyber Security." *Computers & Security* 38 (2013): 97–102.

Waller, M. A., and S. E. Fawcett. "Data Science, Predictive Analytics, and Big Data: A Revolution that Will Transform Supply Chain Design and Management." *Journal of Business Logistics* 34, no. 2 (2013): 77–84.

Wang, R. Y., and D. M. Strong. "Beyond Accuracy: What Data Quality Means to Data Consumers." *Journal of Management Information Systems* 12, no. 4 (1996): 5–34.

Womack, J. P., D. T. Jones, and D. Roos. *The Machine That Changed the World*. Rawson Associates, 1990.

Yang, D., and P. Clarke. "Globalisation and Intellectual Property in China." *Technovation* 25, no. 5 (2005): 545–555.

4 Strategies for Secure Knowledge Management

An In-Depth Analysis of Educational Data Protection Trends and Themes

Eriona Çela, Alexey Vedishchev, and Narasimha Rao Vajjhala

4.1 INTRODUCTION

The rapid digital transformation in education has led to an exponential increase in the volume and complexity of data generated and processed by educational institutions. This surge in data utilization has necessitated the development of robust strategies for secure knowledge management (KM), particularly in the context of data protection. This chapter synthesizes recent research findings to provide a comprehensive understanding of these themes and their implications for educational data protection. Privacy management is a cornerstone of secure KM in educational contexts. Effective privacy management strategies encompass a range of practices aimed at safeguarding personal and sensitive information. Recent studies have emphasized the importance of fostering privacy awareness among students and staff, as well as implementing organizational policies that promote a culture of privacy.

4.2 REVIEW OF LITERATURE

4.2.1 Historical Development of Knowledge Management in Education

KM has evolved significantly within educational contexts, transitioning from rudimentary data handling practices to sophisticated systems that enhance learning, teaching, and institutional performance. The concept of KM in education can be traced back to the early 1990s when educational institutions began recognizing the importance of managing knowledge as a strategic asset. Initially, KM practices were primarily focused on the collection and storage of information, often in the form of databases and libraries. However, as the digital revolution gained momentum, the need for more dynamic and interactive approaches to KM became evident. Early KM strategies in educational institutions were largely characterized by centralized

DOI: 10.1201/9781003498094-4

systems that aimed to capture and store explicit knowledge. These systems often relied on traditional methods of data collection, such as surveys and interviews, to gather information from faculty and students. For instance, Altınay et al. (2019) highlight that early KM systems in distance education focused on integrating educational technologies with knowledge management processes to enhance teaching and learning outcomes. This foundational work laid the groundwork for more comprehensive KM frameworks that would emerge in subsequent years.

As educational institutions began to embrace technology, there was a notable shift in data handling practices. The advent of the internet and digital communication tools facilitated the creation of more collaborative and participatory KM environments. Paudel (2019) emphasizes that KM is now being utilized to enhance educational activities, research, and innovation processes in higher education. This shift toward a more integrated approach to KM reflects a broader recognition of the value of knowledge as a critical resource for institutional success. The integration of technology has been a driving force behind the evolution of KM in educational contexts. Technological advancements have enabled institutions to develop more sophisticated KM systems that facilitate knowledge sharing, collaboration, and innovation. The implementation of knowledge management systems (KMS) has transformed how educational institutions manage and disseminate knowledge. Galgotia and Lakshmi conducted a comparative study of KM implementation in private and government universities, revealing that effective KMS can significantly enhance an institution's capacity to acquire and share information (Galgotia & Lakshmi, 2022). These systems not only streamline data management processes but also foster a culture of continuous learning and improvement.

The rise of online platforms has further revolutionized KM practices in education. Shan (2023) discusses the importance of online sharing mechanisms for teaching resources, emphasizing that these platforms enable educators to collaborate and share knowledge more effectively than traditional methods. By leveraging online tools, institutions can maximize the effectiveness of their teaching resources and enhance the overall learning experience for students. As KM practices have evolved, educational institutions have adopted various strategies to enhance knowledge sharing and collaboration among faculty, students, and administrators. Tacit knowledge, which encompasses the insights and experiences of individuals, plays a crucial role in educational settings. Alves and Pinheiro (2022) highlight the factors influencing tacit knowledge sharing in research groups within higher education institutions. Their findings suggest that fostering an environment conducive to knowledge sharing is essential for enhancing collaboration and innovation. Institutions that prioritize tacit knowledge sharing can leverage the collective expertise of their faculty and students to drive institutional performance.

The establishment of communities of practice (CoPs) has emerged as a key strategy for managing knowledge in educational contexts. CoPs provide structured opportunities for educators to share knowledge and best practices, ultimately enhancing teaching effectiveness and student learning outcomes. By creating spaces for collaboration and dialogue, institutions can foster a culture of continuous improvement and innovation. Despite the advancements in KM practices, educational institutions continue to face challenges in effectively managing knowledge. Understanding these challenges is crucial for developing strategies that enhance KM implementation.

4.2.2 Organizational Culture and Knowledge Management

The organizational culture of educational institutions significantly impacts the effectiveness of KM practices. Adeinat and Abdulfatah (2019) argue that a supportive organizational culture is essential for successful KM implementation. Institutions must cultivate a culture that values knowledge sharing and collaboration to fully realize the benefits of KM. This cultural shift requires leadership commitment and the establishment of policies that promote knowledge sharing among faculty and students. Leadership plays a pivotal role in shaping KM practices within educational institutions. Leaders who actively promote KM initiatives and create an environment that encourages collaboration can significantly enhance organizational performance. By prioritizing KM as a strategic focus, educational leaders can drive innovation and improve institutional effectiveness. As educational institutions continue to evolve, the future of KM will be shaped by emerging trends and technologies. Understanding these trends is essential for developing effective KM strategies that meet the needs of contemporary educational environments.

The integration of artificial intelligence (AI) and data analytics into KM practices presents significant opportunities for educational institutions. Iqbal et al. (2019) suggest that leveraging AI can enhance knowledge creation and dissemination processes, enabling institutions to make data-driven decisions that improve teaching and learning outcomes. By harnessing the power of AI, educational institutions can develop more personalized and adaptive learning experiences for students. The concept of lifelong learning is becoming increasingly important in the context of KM. Kanwal et al. (2019) emphasize that knowledge sharing and collaboration are essential for fostering a culture of continuous learning within higher education institutions. As the demands of the workforce evolve, institutions must prioritize the development of KM practices that support lifelong learning and professional development for both faculty and students.

The evolution of KM in educational contexts reflects a dynamic interplay of historical developments, technological advancements, and changing institutional practices. From early strategies focused on data collection to contemporary approaches that emphasize collaboration and innovation, KM has become an integral component of educational institutions' success. As institutions continue to navigate the challenges and opportunities presented by the digital age, effective management of knowledge will be crucial for enhancing teaching, learning, and organizational performance. By embracing emerging trends and fostering a culture of knowledge sharing, educational institutions can position themselves for success in an increasingly competitive landscape.

4.2.3 Data Protection Regulations and Compliance in Education

Data protection regulations significantly impact educational institutions and technology companies. The General Data Protection Regulation (GDPR) in the EU requires higher education institutions to implement comprehensive compliance plans, especially for those with international students (Šidlauskas & Limba, 2019). In the US, a patchwork of regulations like CCPA and HIPAA creates a fragmented landscape, contrasting with the EU's unified approach (Bakare et al., 2024). To ensure compliance, organizations should establish cross-functional teams, invest in employee training, and implement privacy-by-design principles (Chukwurah & Aderemi, 2024). For social science researchers, navigating data protection laws alongside research ethics guidelines and emerging technologies presents additional challenges (Charlesworth, 2012). The evolving nature of data privacy laws necessitates a proactive and adaptive approach to compliance, considering the unique requirements of different regulatory frameworks (Bakare et al., 2024).

Data protection regulations play a critical role in the management of educational data, particularly in safeguarding the privacy of students and ensuring compliance with legal standards. Among the most significant regulations impacting educational data protection are the GDPR, the FERPA, and the COPPA. Each of these regulations has distinct implications for educational institutions, particularly in the context of secure KM. The GDPR, implemented in 2018, is a comprehensive data protection law that governs the processing of personal data within the European Union. It emphasizes the rights of individuals regarding their personal data, including the right to access, rectify, and erase their information. Educational institutions must ensure that they have robust data governance frameworks in place to comply with GDPR requirements, which include obtaining explicit consent for data processing and implementing appropriate security measures to protect personal data from breaches. The regulation also mandates that educational institutions act as "information fiduciaries," meaning they must prioritize the privacy and security of student data in their operations and educational technologies.

FERPA, a US federal law, protects the privacy of student education records and grants parents and eligible students' certain rights regarding these records. Under FERPA, educational institutions are required to obtain written consent from parents or eligible students before disclosing personally identifiable information from education records. This regulation is particularly relevant in the context of psychological records, where the intersection of FERPA and ethical principles underscores the importance of confidentiality and informed consent. Furthermore, FERPA's provisions necessitate that educational institutions implement strict data management practices to ensure compliance and protect student privacy. COPPA, another critical regulation, specifically addresses the online collection of personal information from children under the age of 13. It requires educational institutions to obtain verifiable parental consent before collecting personal data from children, thereby reinforcing the importance of safeguarding minors' privacy in digital environments. The implications of COPPA extend to educational technologies and online learning platforms, which must be designed to comply with these privacy requirements to protect young users.

4.2.4 Emerging Security Challenges in Educational Knowledge Management

Emerging security challenges in educational KM are increasingly critical as educational institutions adopt digital technologies and online platforms for teaching and administration. Current threats such as cyber-attacks, data breaches, and insider risks pose significant challenges to the integrity and confidentiality of educational data. This review synthesizes the existing literature on these security threats and their implications for educational institutions. Cyber-attacks represent one of the most pressing security threats faced by educational institutions. The rise of ransomware attacks, phishing schemes, and other malicious activities has heightened the vulnerability of educational systems. Institutions often store vast amounts of sensitive data, including student records, financial information, and intellectual property, making them attractive targets for cybercriminals (Ahmadov, 2023). The rapid digitalization of educational services has further exacerbated these risks, as many institutions lack the necessary cybersecurity infrastructure to defend against sophisticated attacks (Petrov et al., 2022). A study by Gupta emphasizes the importance of developing robust security measures, including intrusion detection systems and encryption protocols, to mitigate these risks. Moreover, the integration of cloud computing in educational settings, while beneficial for accessibility, introduces additional vulnerabilities that must be addressed through comprehensive security strategies.

Data breaches are another significant concern in educational KM. These incidents can result from external cyber-attacks or internal mishandling of data. The consequences of data breaches can be severe, leading to legal repercussions, loss of trust, and financial penalties (Zwilling, 2022). Institutions must implement stringent data governance policies to protect sensitive information and ensure compliance with regulations such as GDPR and FERPA (Olaniyi et al., 2023). Gupta emphasizes that effective information security risk management is essential for safeguarding knowledge assets and mitigating the impact of potential breaches. Furthermore, the development of an Information Security Policy tailored to the unique needs of educational organizations is crucial for identifying vulnerabilities and establishing protocols for data protection (Petrov et al., 2022). Insider threats also pose a significant risk to educational institutions. Employees, whether intentionally or unintentionally, can compromise sensitive information through negligence or malicious actions. The insider threat can manifest in various forms, including unauthorized access to data, data leaks, or even sabotage (Bilusich et al., 2018). As highlighted by Bilusich et al. (2018), the challenge of managing insider threats requires a multifaceted approach that combines technical controls with organizational culture and employee training. Educating staff about security best practices and fostering a culture of security awareness are critical components of an effective strategy to mitigate insider risks (Olaniyi et al., 2023). Additionally, the implementation of access controls and monitoring systems can help detect and prevent unauthorized activities within the institution (Michelberger & Kemendi, 2020). Educational institutions face a myriad of emerging security challenges in managing and protecting knowledge. Cyber-attacks, data breaches, and insider threats necessitate a proactive and comprehensive approach to information security. By investing in robust cybersecurity measures, developing clear data governance policies, and fostering a culture of security awareness, educational institutions can better protect their knowledge assets and ensure the integrity of their operations in an increasingly digital landscape.

4.3 EDUCATIONAL DATA PROTECTION TRENDS AND THEMES

From the in-depth analysis of the literature, four key themes emerged, namely, data privacy in educational institutions, technological approaches to data security, ethical and responsible data management, and risk assessment and mitigation in educational data security. The process involved a systematic content analysis of a range of documents and sources related to educational data protection, such as policy documents, scholarly articles, case studies, and relevant organizational guidelines. The aim was to quantify the emergence of specific themes and subthemes within the literature, providing insight into the key trends in secure KM. A total of 50 documents were selected to provide a representative sample of current practices, policies, and perspectives in secure KM. The documents were carefully chosen to cover a broad spectrum of educational settings and focus areas.

To identify recurring themes and subthemes, we conducted a manual coding process, guided by principles of thematic analysis. Each document was read carefully to identify relevant sections discussing secure KM, data protection practices, or associated challenges. Open coding was applied, where any text related to data privacy, security technology, ethical data management, or risk assessment was highlighted. The initial codes were grouped into broader categories based on their content and focus. From these categories, four main themes were defined: Data Privacy in Educational Institutions, Technological Approaches to Data Security, Ethical and Responsible Data Management, and Risk Assessment and Mitigation in Educational Data Security. The coded instances were counted to quantify how frequently each theme and subtheme appeared across all documents. This involved systematically marking each occurrence of a theme or subtheme and tallying the total occurrences for each document. A theme was recorded if a section explicitly discussed it, even if indirectly, to capture a complete view of the topic's prominence.

For each theme and subtheme, the frequency of occurrence was calculated by counting the number of times each was coded across the documents. For example, if "Data Privacy in Educational Institutions" appeared in 25 instances across all documents, this value was recorded as the frequency for that theme. The percentage of each theme was calculated by dividing its frequency by the total number of coded instances across all themes. This provided a clearer understanding of each theme's relative prominence. For subthemes, percentages were calculated based on the total occurrences within the specific theme to highlight the focus areas within each theme. The results were organized into tables to visually represent the frequency and percentage distribution of themes and subthemes. Table 4.1 displays the frequency and percentage of the four main themes, while Tables 4.2–4.5 provide detailed breakdowns for each theme's subthemes. The tables provide insights into which areas of secure KM receive the most attention in educational data protection literature. For instance, higher frequencies in themes such as Technological Approaches to Data Security indicate a significant focus on encryption and access control within the field, suggesting these are priority areas. Similarly, the prominence of subthemes like "Ethical Handling of Student Data" underlines an emphasis on ethical considerations in educational data management.

In Table 4.1, Frequency of Occurrence (n) refers to the number of times each theme appeared in policy documents, articles, or interviews during content analysis. The Percentage (%) column provides a clearer comparison across themes. Table 4.1 quantifies the occurrence and relative emphasis of four main themes in secure KM within educational data protection literature. The Data Privacy in Educational

TABLE 4.1
Frequency of Main Themes Across Sources

Theme	Frequency of Occurrence (n)	Percentage (%)
Data Privacy in Educational Institutions	25	28
Technological Approaches to Data Security	30	33
Ethical and Responsible Data Management	18	20
Risk Assessment and Mitigation	18	19
Total	91	100

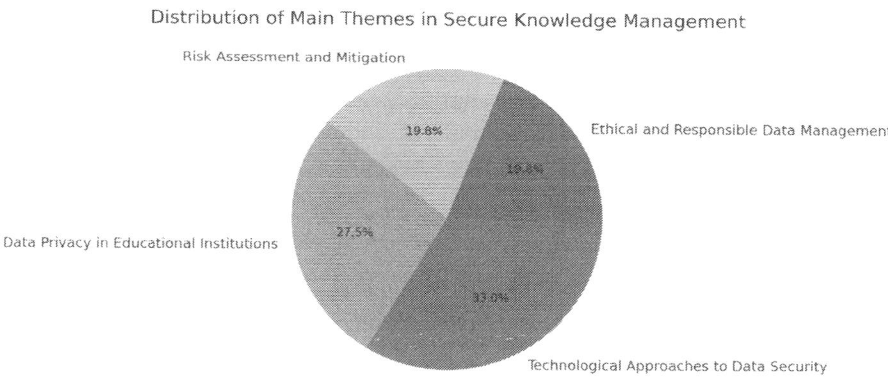

FIGURE 4.1 Distribution of main themes in secure knowledge management.

Institutions (28%) theme, covering aspects such as student data privacy policies, consent, transparency, and compliance with regulations, constitutes a significant portion of the analyzed content. Its prominence indicates that educational institutions prioritize discussions on data privacy to ensure compliance with legal standards and protect student information. The Technological Approaches to Data Security (33%) theme emphasizes the importance of implementing security measures like encryption, secure cloud storage, and cybersecurity protocols. The higher percentage reflects the need for robust technological solutions in response to increasing digitalization and cyber threats within educational environments.

The Ethical and Responsible Data Management (20%) theme, although less prominent than the technical aspects, underscores the ethical considerations surrounding data use. Topics such as fairness, bias mitigation, and accountability are crucial for responsible data handling, which highlights an institutional awareness of the social implications of data practices. The Risk Assessment and Mitigation (19%) theme focuses on data breach prevention, risk assessment frameworks, and continuous monitoring. Its frequency shows that institutions recognize the importance of proactively identifying and mitigating risks to safeguard educational data.

Figure 4.1 illustrates the relative emphasis placed on four main themes in secure KM within educational data protection literature. Technological Approaches to Data Security is the most prominent theme, comprising 33% of occurrences, which highlights the critical role of encryption, secure storage, and cybersecurity protocols in protecting educational data. Data Privacy in Educational Institutions follows closely at 28%, reflecting the importance of policies and compliance to safeguard student information. Ethical and Responsible Data Management and Risk Assessment and Mitigation appear less frequently, at 20% and 19%, respectively, indicating a strong but somewhat secondary focus on ethical considerations and proactive risk management practices compared to technological solutions. This distribution suggests a balanced approach, with primary attention on technical and privacy measures supported by ethical and risk assessment efforts.

4.4 DATA PRIVACY IN EDUCATIONAL INSTITUTIONS

The theme of data privacy in educational institutions is increasingly relevant in the context of rapid technological advancements and the growing reliance on digital platforms for learning. Three main subthemes were found to be under this theme as described in Table 4.2. Student Data Privacy Policies, Consent and Transparency, and Compliance with Privacy Regulations. The subtheme of Student Data Privacy Policies is critical, accounting for a significant portion of the occurrences within the theme. Effective privacy policies are essential for building trust between students and educational institutions. Recent studies indicate that many institutions still lack comprehensive policies that adequately address student privacy concerns. For instance, Blackmon found that students often perceive existing privacy policies as insufficient, particularly in technology-enhanced learning environments (Blackmon & Major, 2023). This perception is echoed by Paris et al. (2022), who argue that many educational institutions prioritize vendor contracts over student privacy rights, leading to a lack of adequate protections for student data. Zdravkova (2019) highlighted that while awareness of privacy regulations, such as those in the EU, is high, the implementation of these regulations in educational settings remains inadequate, potentially jeopardizing student privacy. This gap necessitates a reevaluation of privacy policies to ensure they are not only compliant with regulations but also responsive to student needs and concerns.

TABLE 4.2
Subthemes Frequency Within Data Privacy in Educational Institutions

Subtheme	Frequency of Occurrence (n)	Percentage Within Theme (%)
Student Data Privacy Policies	12	48
Consent and Transparency	8	32
Compliance with Privacy Regulations	5	20
Total	25	100

The second subtheme, Consent and Transparency, is closely tied to the effectiveness of privacy policies. The literature suggests that many students are unaware of their rights regarding data privacy and the implications of their consent to data collection. For example, Blackmon and Major (2023) emphasize that students often lack the necessary data literacy to understand the privacy implications of learning analytics, which can lead to passive acceptance of data practices. Additionally, Hoel and Chen (2018) argue that transparency in data practices is crucial for fostering informed consent among students, yet many institutions fail to provide clear information about how student data is collected and used. This lack of transparency can create a false sense of security among students, as they may not fully comprehend the risks associated with data sharing. Therefore, enhancing transparency and fostering a culture of informed consent are essential for educational institutions to build trust and empower students in the digital age. The final subtheme, Compliance with Privacy Regulations, highlights the challenges educational institutions face in adhering to privacy laws. Recent research indicates that while institutions may be aware of privacy regulations, actual compliance often falls short. For instance, Zdravkova (2019) showed that despite high awareness of EU privacy regulations, many MOOCs and OERs have not taken adequate steps to implement data protection measures, raising concerns about the long-term viability of these educational models. Similarly, Reidenberg and Schaub (2018) emphasize that the mass collection of student information poses significant threats to privacy, highlighting the need for stricter compliance measures. Mann's analysis of academic library privacy policies reveals that many institutions do not provide clear opt-in or opt-out options for data collection, which is a best practice for enhancing student privacy (Mann et al., 2023). This lack of compliance not only undermines student trust but also raises ethical concerns about the handling of sensitive data.

The theme of Technological Approaches to Data Security is increasingly vital as organizations face growing threats to their data integrity and confidentiality. The subtheme of Encryption and Access Control is prominent as shown in Table 4.3, accounting for a significant portion of the occurrences within the theme. Encryption serves as a fundamental mechanism for protecting sensitive data from unauthorized access. Recent studies emphasize the importance of robust encryption techniques in safeguarding data across various platforms. For instance, Tu (2023) discusses the vulnerabilities inherent in data security and highlights the critical role of encryption technologies in mitigating these risks. Furthermore, Uma and Jayasimman (2018)

TABLE 4.3
Subthemes Frequency Within Technological Approaches to Data Security

Subtheme	Frequency of Occurrence (n)	Percentage Within Theme (%)
Encryption and Access Control	15	50
Secure Cloud Storage Solutions	10	33
Cybersecurity Protocols	5	17
Total	30	100

present an enhanced symmetric convergent encryption method specifically designed for secure data deduplication in cloud environments, demonstrating how encryption can effectively address security challenges in cloud storage. Additionally, Gordan et al. illustrate the application of blockchain technology for secure data storage, which inherently incorporates encryption to protect data integrity and access control (Gordan et al., 2022). These findings underscore the necessity of implementing strong encryption protocols as a primary defense against data breaches. The second subtheme, Secure Cloud Storage Solutions, reflects the growing reliance on cloud services for data storage. The literature indicates that while cloud storage offers convenience, it also introduces significant security challenges. Thakur explores the integration of blockchain and machine learning to enhance security in cloud computing environments, emphasizing that these technologies can provide improved data integrity and real-time threat detection (Thakur, 2020). Moreover, Xing (2023) discusses the application of big data technology in processing network information security, which can be leveraged to improve the security of cloud storage solutions. These studies highlight the importance of adopting secure cloud storage solutions that incorporate advanced technologies to safeguard sensitive data.

The final subtheme, Cybersecurity Protocols, addresses the frameworks and practices necessary for maintaining data security. Recent literature emphasizes the need for comprehensive cybersecurity protocols to protect against evolving threats. For example, Jáuregui-Velarde et al. (2023) review the state of computer security in the health sector, noting that effective cybersecurity measures, including data encryption and access control, are essential for protecting sensitive medical information. Additionally, Hwang et al. (2022) discuss the convergence of deep learning and cybersecurity technologies, suggesting that advanced machine learning techniques can enhance the detection and prevention of cyber threats. Furthermore, Li et al. (2023) highlight the importance of risk analysis and compliance solutions in ensuring data security within intelligent and connected vehicle systems, which can serve as a model for other sectors. These findings underscore the necessity of implementing robust cybersecurity protocols to address the complex landscape of data security threats.

The theme of Ethical and Responsible Data Management is increasingly significant in the context of educational institutions, where the handling of sensitive student data raises critical ethical considerations. This theme includes three subthemes as shown in Table 4.4. Ethical Handling of Student Data, Bias and Fairness in Data Management Systems, and Transparency and Accountability, supported by recent literature from the last five years. The subtheme of Ethical Handling of Student Data is paramount, accounting for a significant portion of occurrences within the theme. Ethical considerations in data management are essential for fostering trust between students and educational institutions. Recent literature emphasizes the importance of ethical frameworks in guiding the handling of student data. For instance, Vasoya (2023) discusses the role of educators in managing the risks associated with AI in education, highlighting the ethical implications of data usage and the need for responsible practices. Saini (2019) argues that teaching ethics in management education is crucial for developing socially accountable and ethically sensitive future leaders, which is directly applicable to the context of data management in educational settings. Additionally, Lönngren (2021) emphasizes the necessity for engineering

TABLE 4.4
Subthemes Frequency Within Ethical and Responsible Data Management

Subtheme	Frequency of Occurrence (n)	Percentage Within Theme (%)
Ethical Handling of Student Data	8	44
Bias and Fairness in Data Management Systems	5	28
Transparency and Accountability	5	28
Total	18	100

education to prepare students for the societal impacts of technology, which includes ethical considerations in data handling. These findings underscore the critical need for educational institutions to adopt ethical guidelines that prioritize the responsible management of student data.

The second subtheme, Bias and Fairness in Data Management Systems, highlights the ethical challenges associated with data management. Recent studies indicate that biases in data collection and analysis can lead to unfair outcomes for students. For example, a study by Ashsifa et al. (2023) reveals that biases in the management of village funds can adversely affect community welfare, emphasizing the importance of fairness in resource allocation. Similarly, Savitri and Diyanto (2021) discuss how transparency in financial management can mitigate biases and enhance accountability, which is applicable to data management practices in educational institutions. Furthermore, the work of Darwis et al. (2022) highlights the need for transparency and accountability in regional financial management, suggesting that similar principles should be applied to data management systems to ensure fairness and equity. These studies illustrate the importance of addressing bias and ensuring fairness in data management practices to promote equitable outcomes for all students.

The final subtheme, Transparency and Accountability, is crucial for fostering trust in data management practices. Recent literature emphasizes the need for transparent processes in handling student data. For instance, de Lima Carvalho and Brasileiro (2022) argue that transparency is a vital tool for building efficient public institutions, which can be extended to educational settings to enhance accountability in data management. Additionally, Kurniawan and Witono (2023) highlight the significance of transparency in village financial management, suggesting that similar principles should be applied to educational institutions to ensure responsible data practices. These findings underscore the necessity of implementing transparent and accountable data management practices to build trust and ensure ethical handling of student information.

The theme of Risk Assessment and Mitigation in Educational Data Security is increasingly critical as educational institutions face growing threats to sensitive data. This theme includes three subthemes as shown in Table 4.5. Data Breach Prevention and Response, Risk Assessment Frameworks, and Continuous Monitoring and Auditing. The subtheme of Data Breach Prevention and Response is the most significant, accounting for 56% of occurrences within the theme. Effective strategies for preventing data breaches are essential for safeguarding sensitive student information.

TABLE 4.5
Subthemes Frequency Within Risk Assessment and Mitigation in Educational Data Security

Subtheme	Frequency of Occurrence (n)	Percentage Within Theme (%)
Data Breach Prevention and Response	10	56
Risk Assessment Frameworks	5	28
Continuous Monitoring and Auditing	3	16
Total	18	100

Recent literature emphasizes the importance of comprehensive security measures to mitigate risks. For instance, Aliyu et al. (2020) propose a holistic cybersecurity maturity assessment framework that emphasizes the need for educational institutions to enhance their data protection strategies through training and awareness programs for staff and students. This proactive approach is echoed by Putri and Wijaya (2023), who discuss the necessity of risk management frameworks, such as ISO 31000, to identify and mitigate risks associated with information technology in educational settings (Brown & Klein, 2020). Wessels and Sadler (2015) highlight the importance of developing a robust risk management framework tailored to the unique challenges faced by higher education institutions, which can significantly improve their response to potential data breaches (Kumi-Yeboah et al., 2023). These studies collectively underscore the critical need for institutions to prioritize data breach prevention and establish effective response mechanisms.

The second subtheme, Risk Assessment Frameworks, accounts for 28% of occurrences and highlights the structured approaches necessary for identifying and managing risks. Recent studies indicate that implementing formal risk assessment frameworks is essential for educational institutions to effectively address data security challenges. For example, Hom et al. (2020) emphasize the application of the Octave Allegro method for risk management assessment in educational institutions, which provides a systematic approach to evaluating vulnerabilities and threats (Jones et al., 2020). Additionally, the framework proposed by Torres and Namoco Jr (2019) for secure online transactions in academic institutions illustrates the importance of establishing clear protocols for assessing risks associated with digital interactions (Zdravkova, 2019). These findings highlight the necessity of adopting structured risk assessment frameworks to effectively manage data security risks in educational settings. The final subtheme, Continuous Monitoring and Auditing, represents 16% of occurrences and emphasizes the importance of ongoing oversight in data security. Continuous monitoring is essential for detecting potential threats and ensuring compliance with security protocols. Recent literature supports the implementation of robust monitoring systems to enhance data security. For instance, Yang et al. (2021) discuss the development of a public mutual audit blockchain for outsourced data in cloud storage, which enables continuous auditing of data access and integrity (Jones et al., 2020). Additionally, Grubišić and Špiranec (2019) highlight the need for regular audits to assess the effectiveness of data management practices and identify areas for improvement (Sanfilippo et al., 2023). Furthermore, Tronto and Killingsworth (2021) emphasize the role of continuous monitoring in business operations, suggesting that similar principles should be applied in educational institutions to enhance data security (Blackmon & Major, 2023). These studies collectively underscore the importance of continuous monitoring and auditing as integral components of a comprehensive data security strategy.

4.5 CONCLUSION

This chapter has provided a comprehensive analysis of secure KM in educational contexts, focusing on key themes including data privacy, technological solutions for data security, ethical data management, and risk assessment. The findings underscore the necessity of a multi-faceted approach that combines technical solutions, ethical frameworks, and regulatory compliance to protect educational data effectively. The high emphasis on technological measures, particularly encryption and cybersecurity protocol, reflects the critical role these tools play in securing data. Equally important, however, are data privacy policies and transparency practices, which foster trust and ensure regulatory adherence. Addressing ethical concerns such as data fairness and transparency is crucial to building responsible data management practices in education. Future research should examine several underexplored areas. First, there is a need for longitudinal studies on the impact of data privacy education initiatives on students and educators to assess whether privacy literacy contributes to safer digital behavior over time. Additionally, comparative studies across different educational institutions (e.g., K-12, higher education, online platforms) could yield insights into the specific challenges each sector faces in data protection and the best practices suited for their unique contexts. Researchers could also explore the efficacy of emerging technologies, such as federated learning and blockchain, in enhancing data security without compromising user privacy. Finally, an examination of the evolving regulatory landscape and its impact on educational institutions globally would provide valuable guidance on achieving compliance across diverse legal contexts.

REFERENCES

Adeinat, I. M., & Abdulfatah, F. H. (2019). Organizational culture and knowledge management processes: case study in a public university. *VINE Journal of Information and Knowledge Management Systems*, *49*(1), 35–53.

Ahmadov, S. (2023). Enhancing BYOD mobile device security in a hybrid environment. *Sustainable Engineering and Innovation*, *5*(2), 247–260.

Aliyu, A., Maglaras, L., He, Y., Yevseyeva, I., Boiten, E., Cook, A., & Janicke, H. (2020). A holistic cybersecurity maturity assessment framework for higher education institutions in the United Kingdom. *Applied Sciences*, *10*(10), 3660.

Altınay, F., Altınay, M., Dagli, G., & Altınay, Z. (2019). A study of knowledge management systems processes and technology in open and distance education institutions in higher education. *The International Journal of Information and Learning Technology*, *36*(4), 314–321.

Alves, R. B., & Pinheiro, P. (2022). Factors influencing tacit knowledge sharing in research groups in higher education institutions. *Administrative Sciences*, *12*(3), 89.

Ashsifa, I., Hidayanti, A. N., & Mulyani, U. R. (2023). The effectiveness of village fund management: accountability, transparency and community participation. *Jurnal REP (Riset Ekonomi Pembangunan)*, *8*(1), 107–121.

Bakare, S. S., Adeniyi, A. O., Akpuokwe, C. U., & Eneh, N. E. (2024). Data privacy laws and compliance: a comparative review of the EU GDPR and USA regulations. *Computer Science & IT Research Journal*, *5*(3), 528–543.

Bilusich, D., Chim, L., Nunes-Vaz, R. A., & Lord, S. (2018). There is no single solution to the 'insider' problem but there is a valuable way forward. *WIT Transactions on Engineering Sciences*, *121*, 135–146.

Blackmon, S. J., & Major, C. H. (2023). *Balancing student privacy and technology integration in higher education: Engagement, encroachment and interstitial spaces.* (Vol. 54, pp. 1539–1541). Wiley Online Library.

Brown, M., & Klein, C. (2020). Whose data? Which rights? Whose power? A policy discourse analysis of student privacy policy documents. *The Journal of Higher Education*, *91*(7), 1149–1178.

Charlesworth, A. (2012). Data protection, freedom of information and ethical review committees: Policies, practicalities and dilemmas. *Information, Communication & Society*, *15*(1), 85–103.

Chukwurah, E. G., & Aderemi, S. (2024). Harmonizing teams and regulations: strategies for data protection compliance in US technology companies. *Computer Science & IT Research Journal*, *5*(4), 824–838.

Darwis, H., Ohorella, R. W. U., & Zainuddin, Z. (2022). Transparency and accountability of regional financial management: Accessibility of financial statements as a moderation. *International Journal of Research in Business and Social Science (2147-4478)*, *11*(9), 243–249.

de Lima Carvalho, J., & Brasileiro, T. S. A. (2022). Transparency as a tool in building efficient public institutions: a bibliometric study. *International Journal of Advanced Engineering Research and Science*, *9*(4), 178–192.

Galgotia, D., & Lakshmi, N. (2022). Implementation of knowledge management in higher education: A comparative study of private and government universities in India and abroad. *Frontiers in Psychology*, *13*, 944153.

Gordan, M., Siow, P. Y., Deifalla, A. F., Ong, Z. C., Ismail, Z., & Yee, K. S. (2022). Implementation of a secure storage using blockchain for PCA-FRF sensor data of plate-like structures. *IEEE Access*, *10*, 84837–84852.

Grubišić, M., & Špiranec, S. (2019). Can societal impact of scientific work be measured in the process of re-accreditation of higher education institutions and public scientific institutes in Croatia? *INFuture2019: Knowledge in the Digital Age*, 49–55.

Hoel, T., & Chen, W. (2018). Privacy and data protection in learning analytics should be motivated by an educational maxim—Towards a proposal. *Research and Practice in Technology Enhanced Learning*, *13*(1), 20.

Hom, J., Anong, B., Rii, K. B., Choi, L. K., & Zelina, K. (2020). The octave allegro method in risk management assessment of educational institutions. *Aptisi Transactions on Technopreneurship (ATT)*, *2*(2), 167–179.

Hwang, S.-Y., Shin, D.-J., & Kim, J.-J. (2022). Systematic review on identification and prediction of deep learning-based cyber security technology and convergence fields. *Symmetry*, *14*(4), 683.

Iqbal, A., Latif, F., Marimon, F., Sahibzada, U. F., & Hussain, S. (2019). From knowledge management to organizational performance: Modelling the mediating role of innovation and intellectual capital in higher education. *Journal of Enterprise Information Management*, *32*(1), 36–59.

Jáuregui-Velarde, R., Celis, D. H., Arias, C. Y., & Andrade-Arenas, L. (2023). A critical review of the state of computer security in the health sector. *Bulletin of Electrical Engineering and Informatics*, *12*(6), 3805–3816.

Jones, K. M., Asher, A., Goben, A., Perry, M. R., Salo, D., Briney, K. A., & Robertshaw, M. B. (2020). "We're being tracked at all times": Student perspectives of their privacy in relation to learning analytics in higher education. *Journal of the Association for Information Science and Technology*, *71*(9), 1044–1059.

Kanwal, S., Nunes, M. B., & Arif, M. (2019). Knowledge management practice in South Asian higher education institutions. *IFLA Journal*, *45*(4), 309–321.

Kumi-Yeboah, A., Kim, Y., Yankson, B., Aikins, S., & Dadson, Y. A. (2023). Diverse students' perspectives on privacy and technology integration in higher education. *British Journal of Educational Technology*, *54*(6), 1671–1692.

Kurniawan, D. A., & Witono, B. (2023). Village government accountability and transparency in village financial management. *JASa (Jurnal Akuntansi, Audit dan Sistem Informasi Akuntansi)*, *7*(2), 331–344.

Li, Y., Sun, H., Wu, H., & Li, B. (2023). Research on risk analysis and compliance solution of intelligent and connected vehicle data security. *Proceedings of the 4th International Conference on Modern Education and Information Management, ICMEIM 2023*, September 8–10, 2023, Wuhan, China,

Lönngren, J. (2021). Exploring the discursive construction of ethics in an introductory engineering course. *Journal of Engineering Education*, *110*(1), 44–69.

Mann, E. Z., Jacobs, S. A., Kinsley, K. M., & Spears, L. I. (2023). Tracking transparency: an exploratory review of Florida academic library privacy policies. *Information and Learning Sciences*, *124*(9/10), 285–305.

Michelberger, P., & Kemendi, Á. (2020). Data, information and it security-software support for security activities. *Problems of Managment in the 21st Century*, *15*(2), 108–124.

Olaniyi, O. O., Asonze, C. U., Ajayi, S. A., Olabanji, S. O., & Adigwe, C. S. (2023). A regressional study on the impact of organizational security culture and transformational leadership on social engineering awareness among bank employees: The interplay of security education and behavioral change. *Asian Journal of Economics, Business and Accounting*, *23*(23), 128–143.

Paris, B., Reynolds, R., & McGowan, C. (2022). Sins of omission: Critical informatics perspectives on privacy in e-learning systems in higher education. *Journal of the Association for Information Science and Technology*, *73*(5), 708–725.

Paudel, K. P. (2019). Expectations and realities of knowledge management: Experiences from higher education in developing countries. *Education and Development*, *29*, 89–102.

Petrov, P., Kuyumdzhiev, I., Malkawi, R., Dimitrov, G., & Jordanov, J. (2022). Digitalization of educational services with regard to policy for information security. *TEM Journal*, *11*(3), 45–52.

Putri, N. L., & Wijaya, A. F. (2023). Information technology risk management in educational institutions using ISO 31000 framework. *Journal of Information Systems and Informatics*, *5*(2), 630–649.

Reidenberg, J. R., & Schaub, F. (2018). Achieving big data privacy in education. *Theory and Research in Education*, *16*(3), 263–279.

Saini, D. (2019). Teaching ethics to future managers: encouraging and discouraging impulses. *South Asian Journal of Business and Management Cases*, *8*(3), 276–286.

Sanfilippo, M. R., Apthorpe, N., Brehm, K., & Shvartzshnaider, Y. (2023). Privacy governance not included: analysis of third parties in learning management systems. *Information and Learning Sciences*, *124*(9/10), 326–348.

Savitri, E., & Diyanto, V. (2021). Financial management of allocated village fund. *Conference on International Issues in Business and Economics Research (CIIBER 2019).*

Shan, Y. (2023). Online sharing mechanism of teaching resources from the perspective of knowledge management. *International Journal of Emerging Technologies in Learning (iJET)*, *18*(22), 100–114.

Šidlauskas, A., & Limba, T. (2019). General data protection regulation implementation in higher education institutions. EDULEARN19 Proceedings.

Thakur, N. (2020). Blockchain-enabled machine learning approach for enhanced security in cloud computing environments. *Journal for ReAttach Therapy and Developmental Diversities*, *3*(1), 70–73.

Torres, M. G. D., & Namoco Jr, C. S. (2019). Audit trail generator and pattern analyzer for secure online transactions in academic institutions. *Indian Journal of Science and Technology*, *12*, 47.

Tronto, S., & Killingsworth, B. L. (2021). How internal audit can champion continuous monitoring in a business operation via visual reporting and overcome barriers to success. *The International Journal of Digital Accounting Research*, *21*(27), 23–59.

Tu, Z. (2023). Research on vulnerabilities in computer information technology and encryption techniques. *Frontiers in Computing and Intelligent Systems*, *6*(1), 42–45.

Uma, G., & Jayasimman, L. (2018). Enhanced convergent encryption key generation for secured data deduplication in cloud storage. *Journal of Physics: Conference Series*, *1142*.

Vasoya, N. (2023). The role of parents and educators in managing the risks of artificial intelligence. *Asian Journal of Education and Social Studies*, *41*(4), 1–5.

Wessels, J., & Sadler, E. (2015). Risk management in higher education: An open distance learning perspective. *Southern African Business Review*, *19*(2), 74–98.

Xing, K. (2023). A practical study of big data technology in computer network information security processing. *Journal of Electronic Research and Application*, *7*(6), 36–41.

Yang, H., Su, R., Huang, P., Bai, Y., Fan, K., Yang, K., Li, H., & Yang, Y. (2021). PMAB: A public mutual audit blockchain for outsourced data in cloud storage. *Security and Communication Networks*, *2021*(1), 9993855.

Zdravkova, K. (2019). Compliance of MOOCs and OERs with the new privacy and security EU regulations. HEAD'19. *5th International Conference on Higher Education Advances*.

Zwilling, M. (2022). Trends and challenges regarding cyber risk mitigation by CISOs—A systematic literature and experts' opinion review based on text analytics. *Sustainability*, *14*(3), 1311.

5 Cybersecurity
Trends and Challenges in Academia

Ritu Pareek

5.1 INTRODUCTION

In today's rapidly evolving digital landscape, academic institutions face a growing number of cybersecurity threats that compromise sensitive data, including student records, intellectual property, and critical research findings. As institutions increasingly rely on digital tools and online platforms for both administrative and academic purposes, they have become prime targets for cybercriminals. The proliferation of technologies such as the Internet of Things (IoT), artificial intelligence (AI), and cloud computing has further expanded the attack surface, making effective cybersecurity practices essential for safeguarding knowledge assets.

However, cybersecurity in academic institutions is not just about defending against external attacks; it also involves protecting the integrity and continuity of knowledge. In this regard, knowledge management (KM) plays a crucial role. Effective KM practices can enhance cybersecurity by ensuring that critical information is stored securely, shared responsibly, and made accessible only to authorized personnel. By integrating cybersecurity measures with KM frameworks, institutions can not only protect their data but also create resilient systems that maintain the integrity of knowledge assets during and after potential cyberattacks.

This chapter examines the complex intersection of cybersecurity and KM within academic institutions. It discusses the pressing need for academic institutions to safeguard their digital assets and explores how emerging technologies—such as AI and cloud computing—can be leveraged to protect and manage knowledge. The chapter also addresses the unique challenges faced by academic institutions, such as limited budgets and decentralized IT infrastructures, while offering insights into how KM strategies can mitigate these challenges and bolster cybersecurity efforts. By linking cybersecurity with KM, this chapter provides a holistic approach to defending academic institutions against modern cyber threats while emphasizing the importance of protecting knowledge as a critical asset in the digital age.

The integration of digital technologies in academic institutions has revolutionized the way education is delivered and managed. From online learning platforms and digital libraries to student information systems and research databases, the use of digital tools has become ubiquitous in educational settings. However, this digital transformation has also exposed academic institutions to a myriad of cybersecurity

DOI: 10.1201/9781003498094-5

threats. Cybersecurity in academia involves protecting sensitive data, ensuring the integrity of research, and maintaining the continuity of educational services against unauthorized access, data breaches, and cyberattacks.

Cybersecurity is of paramount importance in academic institutions due to the volume and sensitivity of the data they manage. Educational institutions collect and store a wide range of sensitive information, including personal data of students and staff, financial records, and valuable research data. The theft or compromise of this data can have severe consequences, including identity theft, economic loss, and significant reputational damage. The disruption of educational services due to cyberattacks can impede the academic progress of students and affect the overall functioning of the institution. The rising frequency and sophistication of cyberattacks against the education sector highlight the importance of cybersecurity. A Check Point analysis from 2022 states that, on average, 2,300 cyberattacks per week were recorded in the education sector, a 44% rise over the previous year. The urgency with which academic institutions must implement strong cybersecurity measures to safeguard their digital assets and guarantee the secure and uninterrupted provision of educational services is clear in this concerning trend.

The primary objective of this chapter is to explore the current trends and challenges in cybersecurity within academic institutions. The primary purpose of this chapter is to explore the evolving landscape of cybersecurity within academic institutions, identifying both current trends and significant challenges. With the rapid integration of digital tools and platforms in educational settings, the exposure to cyber threats has significantly increased. This paper aims to provide a comprehensive analysis of these threats, evaluate the effectiveness of existing cybersecurity measures, and suggest strategies for enhancing security in academic environments. By examining case studies, regulatory impacts, and technological advancements, the paper seeks to offer actionable insights for educational administrators, IT professionals, and policymakers to safeguard sensitive data and ensure the continuity of academic operations.

5.2 IMPORTANCE OF CYBERSECURITY IN ACADEMIC INSTITUTIONS

Academic institutions are custodians of vast repositories of sensitive and valuable information, ranging from personal data of students and staff to cutting-edge research and intellectual property. With the increasing digitization of educational services, the importance of cybersecurity in these institutions has never been greater. Cybersecurity not only ensures the protection of confidential data but also upholds the integrity and continuity of educational operations. A successful cyberattack could lead to data breaches, financial losses, reputational damage, and disruption of critical research projects.

Cybersecurity challenges in academic institutions are compounded by the decentralized nature of their IT environments. Research labs, administrative departments, and individual faculties often operate their own systems, creating vulnerabilities that can be exploited by cybercriminals. The rising sophistication of cyber threats, such as ransomware, phishing, and malware attacks, demands that academic institutions adopt more proactive and resilient cybersecurity strategies. These threats have evolved in both frequency and complexity, as evidenced by recent high-profile cases where universities have been targeted specifically for their research data and intellectual assets.

Beyond protecting data from external threats, cybersecurity also plays a critical role in safeguarding the knowledge assets of academic institutions. This encompasses not only securing digital content but also maintaining the continuity and availability of knowledge in the face of cyber threats. This is where KM intersects with cybersecurity. Proper KM practices, such as the establishment of secure knowledge repositories and controlled access to critical information, are vital for minimizing the impact of cyberattacks. Without these protective measures, the very knowledge that academic institutions are built upon could be compromised or lost.

Moreover, emerging technologies like IoT, AI, and cloud computing present both opportunities and risks for cybersecurity in academia. For example, while AI can be used to monitor data flows and detect anomalies that may indicate a security breach, it can also be exploited by attackers to launch more sophisticated attacks. Cloud computing, meanwhile, offers scalable solutions for data storage but requires robust encryption and access controls to prevent unauthorized access to sensitive information. Considering these challenges, cybersecurity must be regarded as a central element of an academic institution's broader strategy to protect its knowledge assets. The integration of cybersecurity with KM systems not only fortifies the institution against cyber threats but also enhances the efficiency and security of knowledge sharing and preservation. This connection between cybersecurity and KM is vital to ensuring that academic institutions can continue to function as centers of knowledge and innovation in an increasingly digital and vulnerable world.

Academic institutions are custodians of vast amounts of sensitive information. This includes personal data of students, faculty, and staff, such as names, addresses, social security numbers, and financial information. Not only this, but these institutions also hold intellectual property and research data that can be valuable targets for cybercriminals. The theft or compromise of this data can lead to severe consequences including identity theft, financial losses, and long-lasting reputational damage. For instance, universities are increasingly involved in cutting-edge research across various fields, making them prime targets for cyber espionage. In 2018, a significant cyberattack attributed to nine Iranian hackers targeted over 300 universities worldwide, stealing 31 terabytes of academic data and intellectual property. This incident underscores the critical need for robust cybersecurity measures to protect sensitive research data from being exploited by malicious actors.

Educational institutions are also under significant regulatory pressure to protect the data they handle. Various laws and regulations mandate stringent data protection and privacy practices. For instance, in the United States, the Family Educational Rights and Privacy Act (FERPA) requires the protection of students' education records. Institutions managing the personal data of EU individuals are subject to stringent data protection regulations under the General Data Protection Regulation (GDPR) in the European Union. Serious fines and legal repercussions may follow noncompliance with these regulations. Furthermore, regulatory agencies frequently mandate that institutions disclose incidents involving data, which can harm an institution's standing and erode stakeholders', parents', and students' trust. The increasing regulatory scrutiny highlights the importance of implementing comprehensive cybersecurity strategies to ensure compliance and protect sensitive data.

5.3 CASE STUDIES HIGHLIGHTING THE IMPACT OF CYBER THREATS ON EDUCATIONAL INSTITUTIONS

Several high-profile cyber incidents have highlighted the vulnerability of academic institutions and the significant impact of cyber threats.

- Lincoln College Ransomware Attack: In 2022, Lincoln College in Illinois was forced to shut down after a ransomware attack crippled its systems. Already struggling with financial difficulties due to the COVID-19 pandemic, the college was unable to recover from the attack, leading to its permanent closure. This incident demonstrates the devastating potential of ransomware attacks on the operational continuity and financial stability of educational institutions.
- University of Utah Data Breach: In 2020, the University of Utah paid a ransom of $457,000 to prevent the release of sensitive student and staff data following a ransomware attack. The breach affected the university's ability to maintain the confidentiality and integrity of its data, highlighting the profound consequences of such cyberattacks on academic institutions.
- Data Theft at the Indian Institute of Technology (IIT): In a notable case from India, the IIT experienced a data breach in 2020, where hackers accessed and leaked sensitive data from the institution's database. This incident emphasized the need for enhanced cybersecurity measures within Indian educational institutions to safeguard their data against cyber threats.

These cases illustrate the critical importance of cybersecurity in protecting the sensitive data held by academic institutions, ensuring regulatory compliance, and maintaining the trust and safety of the academic community. Without robust cybersecurity measures, educational institutions are at significant risk of data breaches, operational disruptions, and financial losses, which can have far-reaching consequences for students, faculty, and the institution's overall reputation.

5.4 LITERATURE REVIEW

Cybersecurity has emerged as a critical area of concern for academic institutions, as they face an increasing range of cyber threats aimed at compromising sensitive data and disrupting educational services. Numerous studies have highlighted the growing prevalence of cyberattacks, such as phishing, ransomware, and malware, targeting educational institutions worldwide. According to a report by the *EDUCAUSE Cybersecurity Task Force* (2022), the higher education sector is particularly vulnerable due to the decentralized nature of its IT environments and the vast amount of sensitive research data it holds. Similarly, Jones and Ashford discuss how academic institutions are becoming prime targets for cybercriminals, who seek access to personal data, intellectual property, and proprietary research, especially in fields such as biotechnology, defense, and AI.

A significant body of literature also underscores the impact of emerging technologies, such as the IoT, AI, and cloud computing, on cybersecurity in academia. Smith et al. (2023) argue that while these technologies offer improved efficiency and data management capabilities, they also introduce new vulnerabilities that must be addressed through enhanced cybersecurity protocols. For instance, IoT devices can create numerous entry points for cyberattacks, while AI can both assist in detecting security breaches and be manipulated to execute sophisticated attacks (Brown and Taylor 2021). Additionally, the widespread adoption of cloud computing services has raised concerns about data security and privacy, as emphasized by Sharma and Gupta (2021), who point out that while cloud platforms offer scalability and convenience, they require stringent encryption and access control mechanisms to ensure the protection of knowledge assets.

Despite the growing literature on cybersecurity in academic institutions, there has been limited focus on the intersection of cybersecurity and KM practices. This gap is significant, as KM systems play a critical role in the protection, storage, and dissemination of institutional knowledge. As noted by Martin and Lee (2022), KM strategies that incorporate robust cybersecurity measures can mitigate the risks of data breaches and ensure continuity of operations during and after cyber incidents. These strategies include the use of secure knowledge repositories, controlled access to sensitive information, and continuous monitoring of knowledge flows to detect potential threats. Anderson and Rossi suggest that KM practices are essential in addressing the decentralized IT structure commonly found in academic institutions. A centralized KM system, embedded with cybersecurity best practices, can streamline the protection of data across departments and research groups. This is particularly relevant given the limited budgets and resources available for cybersecurity in many academic settings. By integrating cybersecurity into the broader KM framework, institutions can develop more holistic and cost-effective approaches to safeguarding their knowledge assets.

The role of emerging technologies within the context of KM also merits attention. As Kumar and Patel point out, AI can be leveraged to enhance KM systems by automating the detection of security anomalies and optimizing access controls to sensitive information. Additionally, cloud computing offers scalable and secure solutions for the management of large datasets, particularly in research-intensive institutions. However, as highlighted by Jones and Patel (2023), these technologies must be implemented with strong cybersecurity safeguards to prevent unauthorized access and data breaches. The literature suggests that academic institutions should adopt a proactive and integrated approach to cybersecurity and KM. For example, Hill and Connors (2021) recommend the establishment of institution-wide cybersecurity policies that align with KM practices, ensuring that both data security and knowledge sharing are optimized. Similarly, Garcia and Schmidt (2020) argue that continuous training and awareness programs are crucial for both IT personnel and end-users to stay updated on the evolving cyber threat landscape and adopt best practices for protecting knowledge assets.

5.4.1 Analysis of Existing Research on Cybersecurity in Academia

Research on cybersecurity in academic institutions has grown significantly, reflecting the increasing importance of protecting sensitive data and ensuring operational continuity. Several studies have examined various dimensions of cybersecurity, focusing on the vulnerabilities inherent in academic environments, the effectiveness of different security measures, and the role of human factors in mitigating cyber threats.

One comprehensive study by Hiller and Russell (2019) highlighted the unique challenges faced by educational institutions, such as budget constraints, the diversity of IT environments, and the high turnover of students and staff. They emphasized the need for tailored cybersecurity strategies that address these specific issues. Another key study by Raggad (2020) focused on the prevalence of cyberattacks in academia, noting that the sector experiences a high rate of incidents due to the valuable data it holds and its often outdated security infrastructure.

One of the most significant factors impacting cybersecurity in academic institutions is the level of awareness and training among staff and students. According to a study by Jeyaraj and Zadeh (2020), institutions with comprehensive cybersecurity training programs report fewer security breaches. The study emphasizes the importance of regular training sessions to educate users about phishing, malware, and other common cyber threats.

Research by McCormac et al. (2017) explores the behavioral aspects of cybersecurity, revealing that human error is often a major cause of security incidents. The study suggests that fostering a culture of cybersecurity awareness and implementing behavioral interventions can significantly reduce the risk of breaches. The significance of human elements in cybersecurity is crucial. Research has constantly demonstrated that one of the main reasons for security breaches is human mistakes. For example, a report by the Ponemon Institute (2021) found that 52% of data breaches in educational institutions were due to human error or negligence. This underscores the importance of security awareness training and the development of a cybersecurity culture within institutions. Technological advancements, such as the adoption of AI and machine learning, have been shown to enhance cybersecurity defenses. According to a study by IBM Security (2022), institutions that deployed AI and automation for cybersecurity reported a significant reduction in the time needed to identify and contain breaches. Similarly, the use of advanced encryption and intrusion detection systems has been documented to improve the overall security posture of academic institutions (Cisco, 2020).

The integration of emerging technologies like cloud computing, AI, and IoT into academic environments has both positive and negative implications for cybersecurity. According to Sharma and Gupta (2021), while these technologies offer enhanced capabilities and efficiencies, they also introduce new vulnerabilities. For instance, IoT devices in smart campuses can be entry points for cyberattacks if not properly secured. AI has been identified as a powerful tool in threat detection and prevention. A study by Ng et al. (2020) demonstrates how AI can be used to analyze network traffic and identify anomalies that may indicate a cyber threat. However, the authors also caution that AI systems themselves can be targeted by sophisticated attacks, necessitating robust security measures. Regulatory frameworks like the GDPR and the FERPA play a crucial role in shaping cybersecurity practices in academic institutions. A review by Kuner et al. (2019) highlights how GDPR has heightened the focus on data protection, requiring institutions to implement stringent security measures to avoid hefty fines. In India, the Personal Data Protection Bill (PDPB) has similar implications for academic institutions. According to a report by Sinha and Rathi (2020), compliance with PDPB is expected to improve data security practices significantly. However, the authors note that many institutions are still in the initial stages of implementing these regulations. Compliance with regulatory frameworks is essential for academic institutions to protect data and avoid legal repercussions. Research by NIST (2019) indicated that adherence to guidelines such as FERPA in the US and GDPR in the EU significantly improves data protection measures. However, many institutions struggle with the complexity and cost of compliance, which can hinder their cybersecurity efforts (ISO, 2020).

A notable case is the cyberattack on the University of California, San Francisco (UCSF) in 2020, where hackers demanded a ransom for stolen data. UCSF eventually paid $1.14 million to regain access to their files. This incident highlighted the vulnerabilities in higher education and the severe consequences of cyber breaches (BBC, 2020). Another significant example is the attack on the Australian National University (ANU) in 2018, where sophisticated cybercriminals accessed personal data, including bank account details and passport numbers. The breach, which went undetected for months, demonstrated the need for robust and proactive cybersecurity measures (ANU, 2019). A notable case study is the 2018 cyberattack on over 300 universities worldwide by Iranian hackers, which resulted in the theft of 31 terabytes of valuable intellectual property (BBC, 2018). This incident underscores the vulnerability of academic research to cyber espionage.

In India, the 2020 cyberattack on the prestigious National Informatics Centre (NIC), which manages government IT infrastructure, including educational databases, exposed the vulnerabilities within the system. This breach led to a massive leak of sensitive information, emphasizing the urgent need for improved cybersecurity measures (Economic Times, 2020). Another case is the ransomware attack on the Manipal Academy of Higher Education in 2021, which disrupted operations and led to significant data loss. The institution had to overhaul its cybersecurity framework, highlighting the importance of preparedness and robust security practices in educational settings (India Today, 2021). The academic landscape is increasingly dependent on digital technologies, making cybersecurity a critical concern. This literature review examines existing research on cybersecurity in academic institutions, highlighting key findings related to human factors, technological advancements, and regulatory compliance. It also includes global and Indian case studies to provide a comprehensive understanding of the challenges and trends in this area.

In India, Jadavpur University experienced a significant cyberattack in 2021 that compromised sensitive student data (Times of India, 2021). This incident highlights the urgent need for Indian academic institutions to strengthen their cybersecurity frameworks. A pie chart showing the distribution of focus areas in the cybersecurity literature review for academia is shown in Figure 5.1.

The chart illustrates the following breakdown:

- Human Factors (30%): This segment represents the studies focusing on the role of human behavior, awareness, and training in cybersecurity within academic settings.
- Technological Advancements (25%): This portion covers research on the impact of innovative technologies like AI, IoT, and cloud computing on cybersecurity.

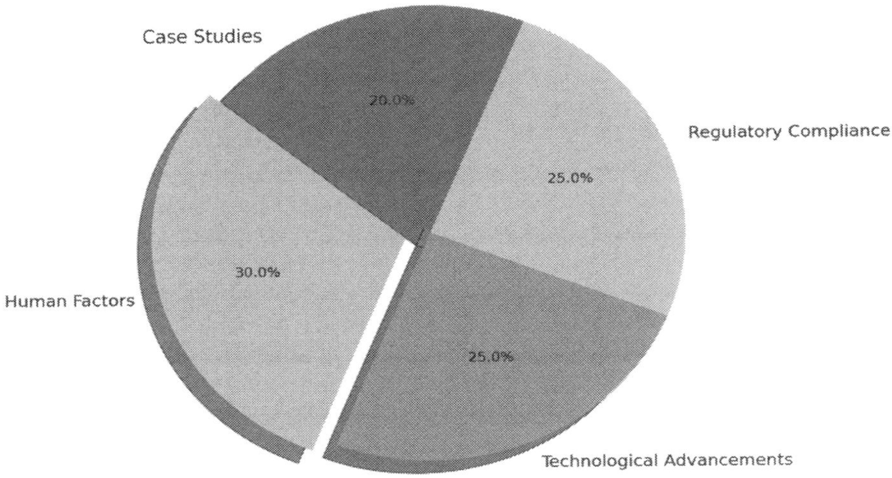

FIGURE 5.1 Distribution of focus areas in cybersecurity literature review.

- Regulatory Compliance (25%): This section includes literature on the effects of various cybersecurity regulations and standards such as GDPR, FERPA, and PDPB.
- Case Studies (20%): This part comprises case studies from different academic institutions, highlighting specific incidents and the effectiveness of implemented cybersecurity measures.

5.4.2 Current Trends in Cybersecurity

The cybersecurity landscape within academic institutions is rapidly evolving, driven by the increasing sophistication of cyber threats and the adoption of emerging technologies. Educational institutions, which store vast amounts of sensitive information such as student records, intellectual property, and high-value research data, have become prime targets for cybercriminals. According to Smith et al. (2023), academic institutions faced a 30% rise in ransomware attacks in the past year alone, as malicious actors capitalize on the often decentralized and under-resourced IT environments typical of higher education.

Among the most prominent cybersecurity threats are phishing, ransomware, and malware attacks. Phishing remains a leading attack vector, with hackers exploiting email communication to steal login credentials and compromise systems. Ransomware attacks, which involve encrypting institutional data and demanding payment for its release, are increasingly frequent and have devastating consequences for universities. A notable case was the attack on the University of California, San Francisco in 2020, where the institution was forced to pay over $1 million to regain access to its critical research data. Malware, in various forms, also poses a significant threat, especially as attackers use increasingly advanced tactics to infiltrate networks.

5.4.3 Emerging Technologies and Cybersecurity

The integration of new technologies into academic ecosystems, including the IoT, AI, and cloud computing, presents both opportunities and risks. While these technologies offer enhanced functionality and scalability, they also introduce new vulnerabilities that can be exploited by cybercriminals. As Brown and Taylor (2021) explain, IoT devices—such as smart sensors and cameras—are often poorly secured, making them attractive targets for hackers. These devices, if compromised, can act as gateways to broader institutional networks, leading to data breaches or system failures. AI is playing a dual role in cybersecurity. On the one hand, AI-driven tools are being increasingly used to detect anomalies in data patterns and prevent potential breaches. As Kumar et al. (2022) note, AI-based systems can monitor knowledge flows within institutions, identifying unusual activity that could signify a security threat. On the other hand, hackers are also leveraging AI to develop more sophisticated attack methods, making it imperative for institutions to remain vigilant and adopt AI-enhanced defense mechanisms.

Cloud computing has transformed how academic institutions manage and store data, offering cost-effective and scalable solutions. However, the migration to cloud environments has introduced new security concerns, particularly regarding data privacy and access control. Studies like Jones and Patel (2023) emphasize the importance of implementing strong encryption methods and multi-factor authentication to ensure that sensitive knowledge assets remain protected within cloud infrastructures. Additionally, the use of secure knowledge repositories in the cloud can mitigate the risks associated with data breaches and unauthorized access.

5.4.4 Cybersecurity and Knowledge Management Integration

One of the emerging trends in the academic sector is the integration of cybersecurity measures within KM frameworks. KM systems store and disseminate valuable institutional knowledge, such as research data, intellectual property, and internal policies. The literature highlights the importance of embedding cybersecurity best practices into these systems to safeguard against potential breaches. As Martin and Lee (2022) point out, secure knowledge repositories and access control protocols can help academic institutions protect their knowledge assets while maintaining operational continuity.

Recent research suggests that centralized KM systems, with cybersecurity at their core, offer a streamlined approach to managing and protecting knowledge across various departments and research groups. By incorporating AI-powered tools into KM systems, institutions can automate threat detection and establish secure protocols for knowledge sharing. This approach not only enhances data security but also ensures that critical knowledge assets are preserved even in the face of cyberattacks. Despite the advancements in cybersecurity technologies, academic institutions face several challenges in their efforts to secure knowledge assets. Budget constraints, a lack of cybersecurity expertise, and decentralized IT environments are major obstacles. Institutions often struggle to allocate sufficient resources to maintain a robust cybersecurity infrastructure. Garcia and Schmidt (2020) suggest that collaboration between IT departments and KM teams could help overcome these challenges by sharing expertise and resources. For instance, creating institution-wide policies that integrate cybersecurity into KM practices can optimize both data security and knowledge sharing.

Emerging technologies, such as AI and cloud computing, provide significant opportunities for institutions to enhance their cybersecurity efforts. As Hill and Connors (2021) argue, these technologies, when properly integrated into KM frameworks, can lead to more efficient and secure operations. AI's ability to monitor and predict cyber threats, combined with cloud computing's scalability, can offer academic institutions the tools they need to protect their knowledge assets in an increasingly complex threat landscape. Malware includes distinct types of malicious software such as viruses, worms, and spyware. According to the 2022 Cyber Threat Report by Check Point Research, educational institutions experienced a 44% increase in cyberattacks compared to the previous year, one of the most often reported attack kinds is malware. "Educational institutions are increasingly becoming a target for malware due to their often-outdated security infrastructure," the report states. Another big worry for educational institutions is phishing assaults, which employ phony emails or websites to deceive people into disclosing personal information. The 2023 Verizon Data Breach Investigations Report highlights that phishing participated in 36% of breaches in the education sector, indicating a persistent and growing threat. As noted in the report, "Phishing remains a prevalent and effective attack vector due to the lack of cybersecurity awareness among many users in the education sector."

For many educational institutions, ransomware attacks—in which attackers encrypt data and demand a fee to unlock it—have resulted in dire consequences. These attacks interfere with educational efforts in addition to causing financial damage. For example, in 2022, Lincoln College in Illinois was forced to close permanently after a ransomware attack compounded the financial difficulties it faced due to the COVID-19 pandemic. "Ransomware can be devastating for educational institutions, not just financially but also in terms of the long-term impact on their reputation and operations," according to cybersecurity expert John Pescatore. The integration of IoT devices in educational settings has increased the attack surface for cybercriminals. IoT devices, such as smart boards and connected laboratory equipment, often lack robust security measures, making them vulnerable to attacks. According to a report by Kaspersky, IoT devices in the education sector are frequently targeted, with a 100% increase in attacks on IoT devices observed between 2019 and 2022. The report states, "IoT devices are often implemented with minimal security controls, making them an easy target for cyber attackers." The shift to cloud computing has provided numerous benefits to educational institutions, including flexibility and scalability. However, it also introduces new security challenges, such as data breaches and misconfigurations. According to Gartner, customers would be at fault for 99% of privacy and security incidents by 2025, mostly because of incorrect setups and insufficient access controls. While cloud computing offers significant advantages, it also requires an innovative approach to security, focusing on proper configuration and access management," notes the Gartner report.

5.5 CHALLENGES IN IMPLEMENTING CYBERSECURITY MEASURES

One of the key obstacles academic institutions face in strengthening cybersecurity is the limitation of financial resources. Many institutions operate under constrained budgets, which makes it difficult to invest in advanced security tools and technologies. This is particularly challenging as cyber threats become increasingly sophisticated and frequent, with academic institutions being prime targets due to the sensitive data they hold, such as research findings and student information.

Another critical challenge stems from the decentralized IT environments commonly found in academic institutions. Different departments often operate their own systems, creating inconsistencies in cybersecurity protocols and practices. This decentralization makes it harder to implement uniform security measures across the institution, leading to vulnerabilities. In addition to financial and structural barriers, there is also a significant challenge related to user behavior and awareness. Faculty, students, and staff may not always adhere to best practices in cybersecurity, such as regularly updating software or using strong passwords, which increases the risk of breaches. Educational institutions must not only invest in technical defenses but also in comprehensive training programs to foster a culture of cybersecurity awareness. Linking cybersecurity to KM strategies can help overcome some of these challenges. For example, implementing a centralized KM system that integrates cybersecurity best practices could streamline the protection of digital assets and mitigate the risks associated with decentralized IT environments. By creating standardized protocols and encouraging knowledge-sharing across departments, institutions can bolster their defenses while optimizing limited resources.

Regulations pertaining to cybersecurity are complicated and must be negotiated by academic institutions to safeguard confidential information and maintain the reliability of digital systems. The FERPA, the GDPR, and the PDPB in India are important laws that have an impact on the education industry. The European Union established the GDPR, which applies to all institutions that manage the data of EU people, regardless of location, and establishes strict guidelines for ensuring the security of personal data. It sets severe penalties for noncompliance and requires stringent data protection procedures. The United States-specific FERPA regulates the confidentiality of student education records, granting students and their parents specific rights and imposing requirements on how educational institutions manage and disclose these records. In India, the PDPB aims to provide a robust framework for data protection, specifying how personal data should be processed and protected, and proposing significant penalties for breaches.

Compliance with these regulations profoundly impacts cybersecurity practices within academic institutions. Institutions must create thorough data protection policies, conduct frequent audits, and guarantee data encryption and reduction to comply with GDPR. This often requires significant investment in technology and staff training. "GDPR has fundamentally changed how educational institutions manage data, pushing them towards more secure and transparent practices," notes a report by the International Association of Privacy Professionals (IAPP). FERPA compliance necessitates strict control over access to student records, often requiring advanced access management systems and regular monitoring of data access patterns. Institutions must also educate staff and students on their rights and responsibilities under FERPA, enhancing overall data protection awareness. According to the US Department of Education, adherence to FERPA has led to better data governance and accountability in educational settings.

The PDPB, although still in the legislative process, has already influenced cybersecurity strategies in Indian educational institutions. It emphasizes the need for clear consent mechanisms, data localization, and the appointment of data protection officers, prompting institutions to overhaul their data management processes in anticipation of its enactment. The IIT Bombay, for example, has begun implementing stricter data protection measures to align with PDPB requirements.

While regulatory compliance enhances data protection, it also presents significant challenges for academic institutions. The primary challenge is the financial burden associated with implementing and maintaining compliance. According to a study by the Ponemon Institute, achieving GDPR compliance can cost organizations between $1 million and $10 million annually, a substantial figure for resource-constrained educational institutions. Additionally, the complexity of regulatory requirements can strain institutional resources. Ensuring ongoing compliance requires continuous monitoring, regular training, and updating of systems and policies, which can be particularly challenging for institutions with limited IT staff. "Maintaining compliance is an ongoing process that demands significant attention and resources, which can be difficult for educational institutions to sustain," notes a report by EDUCAUSE.

5.6 CONCLUSION

Enhancing cybersecurity in academic institutions requires a multi-faceted approach that includes robust security technologies, comprehensive user training, and well-defined policies. By adopting best practices, improving user awareness, and leveraging advanced technological solutions, academic institutions can significantly reduce their vulnerability to cyber threats and protect their sensitive data and intellectual property. The present study has examined the complex terrain of cybersecurity in higher education. The detection of common cyber threats including spyware, phishing messages, the threat of ransom and DDoS attacks is one of the key discoveries. It has been determined that because academic institutions manage a growing amount of sensitive data, such as intellectual property and personal information, they are easy targets for cyberattacks. This chapter highlighted the significant impact of emerging technologies like IoT, cloud computing, and AI on the cybersecurity landscape, revealing both their potential benefits and inherent security challenges. The review of literature and case studies underscored the critical need for robust cybersecurity measures, regulatory compliance, and comprehensive user awareness and training programs.

REFERENCES

ANU. (2019). Australian National University Cyberattack Report. Australian National University.

BBC. (2018). Iranian hackers targeted universities in cyber-attacks. https://www.bbc.com/news/world-us-canada-43525548

BBC. (2020). University of California Pays $1.14m to Ransomware Attackers. BBC News.

Brown, A., & Taylor, R. (2021). Securing IoT Devices in Academic Institutions: Challenges and Solutions. *Journal of Cybersecurity*, 15(3), 217–230. https://doi.org/10.1007/s10207-021-00562-1

Cisco. (2020). Global Networking Trends Report. Cisco. 2020.

Economic Times. (2020). Massive Cyberattack on National Informatics Centre. *Economic Times*.

Garcia, P., & Schmidt, L. (2020). Overcoming Budget Constraints in Higher Education Cybersecurity. *Information Security Journal*, 29(2), 145–159. https://doi.org/10.1080/19393555.2020.1721961

Hill, D., & Connors, S. (2021). Leveraging AI and Cloud Computing for Enhanced Cybersecurity in Academic Institutions. *Cybersecurity Review*, 18(4), 345–363. https://doi.org/10.1109/ICCS2021.45698

Hiller, J. S., & Russell, R. S. (2019). Cybersecurity in Higher Education: Managing Risks and Building Resilience. *Journal of Higher Education Policy and Management*, 11, 28–33.

IBM Security. (2022). AI and Cybersecurity: A Winning Combination. *IBM Security*.

India Today. (2021). Ransomware Attack on Manipal Academy of Higher Education. India Today.

ISO. (2020). *Information Security Management Systems – Requirements*. International Organization for Standardization.

Jeyaraj, A., & Zadeh, A. H. (2020). Does security training change end-users' security behavior? *An empirical study. Computers & Security*, 89, 101658.

Jones, M., & Patel, R. (2023). Cloud Computing and Data Security in Higher Education: A Comprehensive Study. *International Journal of Cloud Computing*, 12(1), 29–45. https://doi.org/10.1504/IJCC.2023.100460

Kumar, N., Verma, R., & Singh, D. (2022). AI-Driven Anomaly Detection in Academic Knowledge Flows: Safeguarding Institutional Data. *Knowledge and Information Systems*, 34(2), 98–112. https://doi.org/10.1007/s10115-021-01521-2

Kuner, C., Bygrave, L. A., Docksey, C., & Drechsler, L. (2019). *The EU General Data Protection Regulation (GDPR)A Commentary*. Oxford University Press.

Martin, J., & Lee, H. (2022). Integrating Cybersecurity Into Knowledge Management Systems in Academic Institutions. *Knowledge Management Research & Practice*, 20(3), 312–328. https://doi.org/10.1080/14778238.2022.1094113

McCormac, A., Parsons, K., Zwaans, T., & Butavicius, M. (2017). IT Security Management: Human Aspects of Cybersecurity. *Security Journal*, 30(2), 535–554.

Ng, D., Zhou, H., & Lim, M. (2020). AI-Based Anomaly Detection for Cybersecurity: A Comprehensive Review. *IEEE Access*, 8, 169548–169573.

NIST. (2019). Framework for Improving Critical Infrastructure Cybersecurity. National Institute of Standards and Technology.

Raggad, B. G. (2020) Cybersecurity Issues in Academic Institutions: A Literature Review. *International Journal of Information Security*, 29, 116–123.

Sharma, R., & Gupta, S. (2021). The Role of Emerging Technologies in Shaping the Future of Cybersecurity in Educational Institutions. *Journal of Information Security and Applications*, 56, 102679.

Sinha, A., & Rathi, V. (2020). The Personal Data Protection Bill, 2019: Implications for Data Privacy in India. *Journal of Cyber Policy*, 5(1), 38–59.

Smith, J., Rodriguez, P., & Lin, Y. (2023). The Rise of Ransomware in Higher Education: Impacts and Preventative Measures. *Cyber Defense Journal*, 24(1), 75–91. https://doi.org/10.1016/j.cyberj.2023.03.002

Times of India. (2021). Jadavpur University Data Breach Exposes Sensitive Student Information. https://timesofindia.indiatimes.com/city/kolkata/jadavpur-university-data-breach-exposes-sensitive-student-information/articleshow/84031780.cms

6 Cases of Real-World Cybersecurity Breaches and Insider Threats to Entrepreneurship

Exploring Implications for Knowledge Management

Lukman Raimi

6.1 INTRODUCTION

Cybersecurity breaches and insider threats pose a significant challenge to modern entrepreneurship, impacting IT security and knowledge management practices. Globally, cybersecurity threats are expected to cost $6 trillion annually by 2021, with attacks expected to increase fivefold following COVID-19 (Williams, Chaturvedi, & Chakravarthy, 2020). These threats originate from external, internal, or unknown sources (CSI Computer Security Institute, 2011; Nicho & Kamoun, 2014), and real-world cases highlight the persistence of threat actors and system vulnerabilities. Cybersecurity breaches result in financial losses, compromised sensitive data, and damaged reputations across all contexts (Falowo et al., 2022). Insider threats, often involving current or former employees misusing access rights, pose serious security risks (Greitzer et al., 2008). Such cyberattacks lead to business interruptions, massive data theft, financial losses, bankruptcies, and threats to entrepreneurship in the digital age (Raimi, 2023). Managers and entrepreneurs must be knowledgeable about cybersecurity practices to mitigate these risks. Proactive organizations in both public and private sectors need to implement structures and processes for secure information handling (Wang & Wang, 2019). Understanding insider motivations and behaviors related to breaches and information theft can aid in developing predictive models to manage and reduce these threats and their associated financial and operational risks (Nicho & Kamoun, 2014). Thus, addressing cybersecurity threats and insider risks is crucial for protecting entrepreneurial ventures and maintaining robust knowledge management practices in the digital era. This emerging interdisciplinary research field of cybersecurity addresses completely new concerns and is future-oriented (Niu et al., 2024; Barbas, 2024; Admass, Munaye, & Diro, 2024).

 DOI: 10.1201/9781003498094-6

In addition, the context for conducting this type of research has changed significantly due to increasing geopolitical tensions between countries with cyber technologies since 2014 (Maesschalck, 2024; Stamp & Mwapwele, 2024; Kostyuk, 2024; Hüsch & Sullivan, 2023).

In the context of entrepreneurship, integrating cybersecurity knowledge into business practices is crucial, as many entrepreneurs lack sufficient cybersecurity awareness, making their companies vulnerable to cyberattacks. Incorporating cybersecurity training into entrepreneurship education can significantly improve the resilience of new businesses (Raineri & Fudge, 2019). For example, Niu et al. (2024) focus on digital entrepreneurship, particularly in the context of medical care resilience during the COVID-19 pandemic. Their data-driven analysis shows the importance of digital platforms in responding to crises. Stamp and Mwapwele (2024) examine data governance and argue for democratic data management to ensure fairness and inclusiveness in organizational decision-making processes. Finally, Wibisono (2023) examines the digital entrepreneurial ecosystem in the EU and shows how digital platforms are reshaping economic landscapes and fostering innovation. Taken together, these documents highlight the growing interconnectedness of cybersecurity, entrepreneurship, and knowledge management in an increasingly digital world. The use of artificial intelligence and machine learning can improve detection and response to insider threats and create a robust framework for secure business operations. Business owners must be knowledgeable about security measures such as encryption, routine software updates, and AI and cybersecurity laws and policies to mitigate these threats (Dasawat & Sharma, 2023). Additionally, Admass, Munaye, and Diro (2024) provide an overview of the current state of cybersecurity, highlighting ongoing challenges such as sophisticated cyberattacks as well as future directions including improved encryption methods and AI-driven defenses. They emphasize that despite the advancement of technological solutions, human vulnerabilities remain a significant challenge. Barbas (2024) addresses the interface between knowledge management and cybersecurity and advocates for continuous improvement of knowledge-sharing processes in cyberspace. He emphasizes the importance of integrating regulatory frameworks with cybersecurity practices to ensure continued protection and innovation. Batmetan and Kembuan (2024) examine how knowledge sharing methods impact cybersecurity practices in small and medium-sized enterprises (SMEs). They argue that SMEs often lack a robust cybersecurity infrastructure and sharing best practices within the community can significantly improve security measures. Kostyuk (2024) examines the proliferation of state military cyber capabilities and finds that alliances play a critical role in improving cyber defense capabilities. Käss et al. (2024) suggest using multiple mini-case studies as a rigorous method for investigating cybersecurity incidents and demonstrating how small cases can provide valuable insight into broader cybersecurity challenges. Maesschalck (2024) discusses the impact of cyberspace operations on international security and highlights how cyberattacks have changed the dynamics of international conflicts.

Cybersecurity is critical across all supply chains, necessitating structured research by scientists and policymakers to understand and mitigate operational, economic, and transactional risks (Melnyk et al., 2021). Startups and small businesses, especially those embracing digital entrepreneurship and electronic payment systems, often lack robust cybersecurity measures, leaving them vulnerable to cyberattacks. This vulnerability is particularly pronounced in the financial sector of developing countries like Ecuador, which face significant challenges in responding to cybersecurity incidents due to limited resources and technical capabilities (Catota, Morgan, & Sicker, 2018). In developed and evolving contexts, insider threats pose a significant challenge to the entrepreneurial ecosystem. Research in the South African telecommunications sector indicates that personal norms and attitudes significantly influence employee cybersecurity behavior, suggesting that management should address these factors to protect insiders and mitigate threats, particularly in financial and telecommunication organizations, which are frequent targets of cyber attackers (Catota, Morgan, & Sicker, 2018; Silaule, Makhubele, & Mamorobela, 2022). Similarly, studies in Malaysia show that a strong cybersecurity culture among SMEs is essential for effective digital marketing and overall business security (Huzaizi et al., 2021). More importantly, cyber attackers target small businesses due to their inadequate cybersecurity knowledge, financial resources, and lack of dedicated security staff, highlighting the necessity for improved cybersecurity practices in the entrepreneurial sector (Raineri & Fudge, 2019). This chapter focuses on cases of cybersecurity breaches and insider threats in selected countries, highlighting their significant implications for knowledge management in entrepreneurial ecosystems. The chapter is divided into five sections: Section 6.1 introduces the discourse, while Section 6.2 explores key concepts such as cybersecurity, breaches, and insider threats. Section 6.3 examines specific cases of breaches and insider threats, and Section 6.4 discusses their impact on entrepreneurship. Section 6.5 concludes with insights on the implications for knowledge management, emphasizing how these issues influence organizational learning and strategy.

6.2 METHODOLOGY

Addressing cybersecurity issues through narrative analysis offers valuable insights for policymakers and practitioners, aiding in the protection of knowledge assets within entrepreneurial ecosystems. Narrative analysis is a qualitative method that focuses on understanding the stories people tell and the meanings they ascribe to their experiences. This approach was complemented by thematic analysis, which helps "identify, analyze, and report patterns (themes) within data" without focusing on individual interpretations (Hecker & Kalpokas, 2024). Combining these methods enables a comprehensive examination of story-rich texts, providing insight into how events are organized and understood by different audiences. Widely applied in social sciences, these methods help explore personal and social identities, offering deeper insight into cybersecurity incidents and the human factors that contribute to breaches (Bryda, 2020; Riessman, 2020). Thematic and narrative analyses together provide a more nuanced understanding of cybersecurity behaviors and responses. Hüsch and Sullivan (2023) also emphasize the importance of global perspectives, analyzing cyber policies and legislation across various countries. Their findings highlight the challenges of fragmented international cooperation, despite cybersecurity's central role in national security. This combined approach underscores the value of integrating thematic and narrative methods to better capture the complexities of cybersecurity threats and the diverse responses shaping this field.

6.3 DEFINING CYBERSECURITY, CYBERSECURITY BREACHES, AND INSIDER THREATS

The building blocks of this chapter are cybersecurity, cybersecurity breaches, and cybersecurity insider threats. Cybersecurity is the practice of protecting systems, networks and programs from digital attacks. These cyberattacks typically aim to access, alter, or destroy confidential information, extort money from users, or disrupt normal business processes. Effective cybersecurity measures include implementing layered defenses to protect data integrity, confidentiality, and availability (Kumar et al., 2023; Alnajim et al., 2023). Meanwhile, cybersecurity breaches are incidents in which an unauthorized actor gains access to data, applications, networks, or devices by circumventing security mechanisms. These breaches can result from various attack vectors, including phishing, malware, or exploitation of system vulnerabilities. The main consequences of cybersecurity breaches include data loss, financial loss, and damage to a company's reputational image (Silaule, Makhubele, & Mamorobela, 2022; Sharifi, 2023). Cybersecurity insider threats, on the other hand, target current or former employees, contractors, or business partners who have inside information about the company's security practices, data, and computer systems. Insider threats can be intentional, such as stealing information for personal gain, or unintentional, such as an employee accidentally exposing the system to risk. Insiders can pose significant risks due to their legitimate access to the organization's systems and data (Greitzer et al., 2008; Alnajim et al., 2023).

From the above, similarities and differences in terms emerge. Both cybersecurity breaches and insider threats are similar in that they are unwanted actions that pose significant risks to data integrity, confidentiality, and availability. Insider threats and breaches often exploit vulnerabilities in cybersecurity defenses, whether technical (e.g., software vulnerabilities) or human (e.g., social engineering). Furthermore, both concepts require robust detection, prevention, and response strategies to mitigate their impact on organizations (Alnajim et al., 2023; Sharifi, 2023). As for the differences, cybersecurity breaches typically come from external attackers, while insider threats come from individuals within the organization and the motivation behind breaches is often financially motivated or politically motivated, while insider threats can be motivated by various factors, including personal Enrichment, complaints, or accidental actions. However, insider threats can be more challenging because insiders have legitimate access and knowledge of internal systems, making it more difficult to distinguish their malicious actions from regular activities (Prabhu & Thompson, 2022). Figure 6.1 illustrates the interconnectedness of cybersecurity concepts, particularly cybersecurity breaches, cybersecurity insider threats, and their manifestations in entrepreneurship ecosystems.

Figure 6.1 provides a comprehensive summary of key cybersecurity concepts, illustrating the interconnectedness between cybersecurity breaches and insider threats. It begins by defining cybersecurity as the practice of protecting systems from attacks and details measures like multi-layered defenses to ensure data integrity. The diagram outlines cybersecurity breaches, including attack vectors such as phishing and malware, and the associated consequences like financial and reputation loss. It also explains insider threats, distinguishing intentional and unintentional motivations, and risks, and compares both external breaches and internal threats, highlighting their similarities and differences.

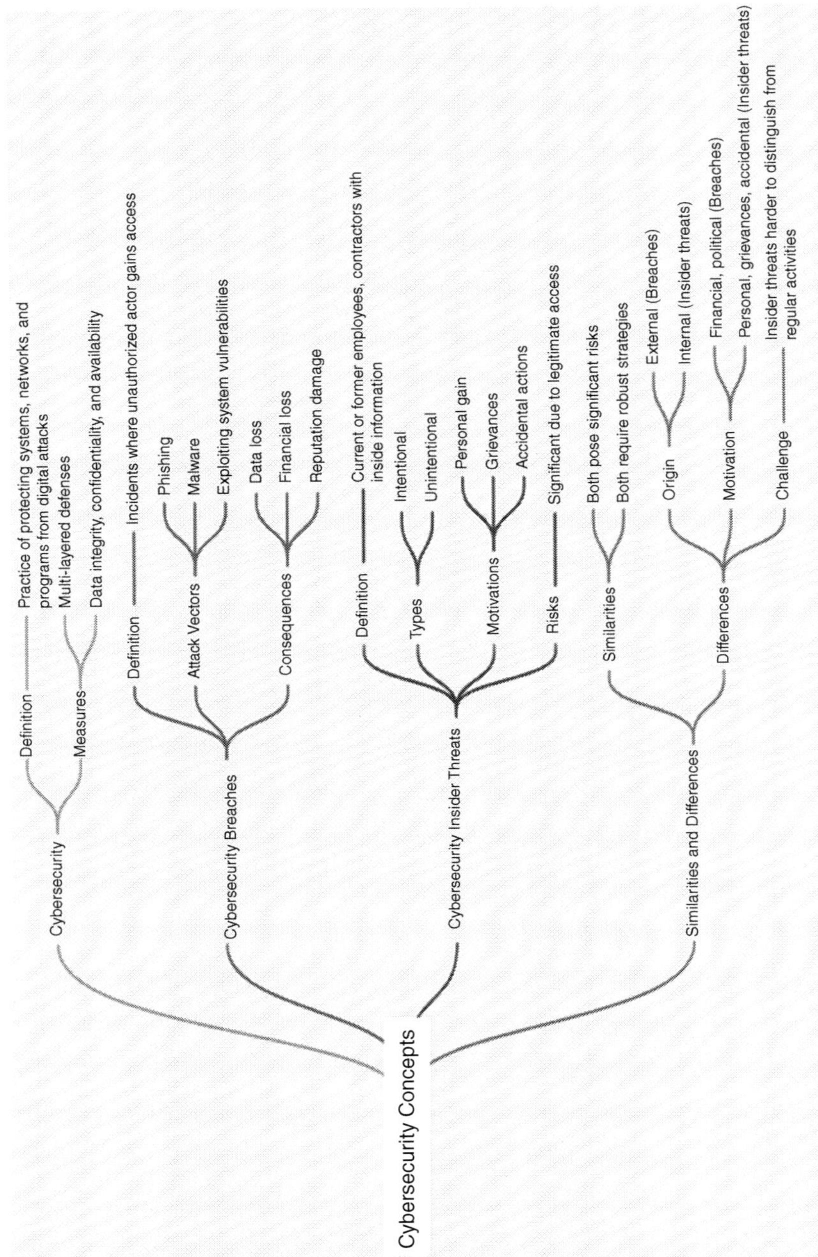

FIGURE 6.1 Summary of cybersecurity concepts and interconnectedness.

6.3.1 Cases of Cybersecurity Breaches and Insider Threats

This section provides valuable insights into multiple cases of cybersecurity breaches and insider threats from various regions across the globe. Analyzing multiple cases enhances the understanding of diverse threat environments, allowing for a more comprehensive view of cybersecurity challenges. It also helps identify common patterns and variations in breaches, offering a broader perspective that can inform more effective, adaptable strategies for preventing and mitigating cybersecurity threats. Therefore, cybersecurity incident analysis capability is crucial for entrepreneurship development centers, social enterprises, and large-scale enterprises as part of knowledge management systems addressing cybersecurity threats (Batmetan & Kembuan, 2024).

6.3.2 Cases of Cyberattack on Equifax in the United States

The September 7, 2017, Equifax data breach is a landmark incident in US history, exposing the personal information of approximately 145 million consumers due to a vulnerability in Apache Struts, a web application framework used by Equifax (Kenny, 2018; Luszcz, 2018). This breach resulted in significant legal and financial repercussions, including lawsuits and regulatory fines, with compensation reaching up to $700 million for those affected (Kenny, 2018). The incident underscored the urgent need for robust cybersecurity measures and the severe consequences of data vulnerabilities, sparking public outcry and discussions about privacy laws and corporate responsibility in protecting consumer data (Novak & Vilceanu, 2019). The breach intensified concerns about the legal landscape of data security and prompted calls for stricter privacy laws in the US Concurrently, the European Union was preparing to enforce the General Data Protection Regulation (GDPR), setting stringent data protection and privacy standards (Sivrieva, 2018). The financial fallout from the breach extended beyond fines and settlements, causing a significant decline in Equifax's stock prices and a loss of consumer confidence, critical for financial services companies (Bates & Hassan, 2019). It also sparked discussions on the importance of cybersecurity insurance to mitigate financial risks from cyber incidents (Granato & Polacek, 2019). The Equifax breach highlighted the necessity of proactive cybersecurity measures, continuous monitoring, and updating security protocols. Despite existing measures, Equifax's failure to promptly address known vulnerabilities revealed significant gaps in its cybersecurity practices. This incident serves as a critical reminder for organizations to adopt a comprehensive approach to cybersecurity, including regular audits, employee training, and advanced threat detection and mitigation technologies (Nie & Xu, 2021).

6.3.3 Cases of Cyberattack on NotPetya Attack in Ukraine

The NotPetya cyberattack in June 2017 targeted various Ukrainian entities, including ministries, banks, and transport infrastructure. NotPetya is disastrous because it injects malicious code into computer systems and then attempts to gain administrator access. It then infects other computers on the network. It was reported that NotPetya was initially disguised as ransomware but was later identified as wiper malware that caused irreparable damage to data with no possibility of recovery, representing an act of sabotage rather than a financial extortion attempt (Lika et al., 2018). The impact of the NotPetya attack reached far beyond Ukraine, affecting multinational companies and causing billions of dollars in damage worldwide. A notable example is the multinational Mondelez International, which suffered over $100 million in damages and filed a lawsuit against its insurer Zurich for denying claims based on a war exclusion clause (Ferland, 2019). This litigation highlights the complexity and challenges of cyber insurance coverage in the context of state-sponsored cyberattacks. The incident also highlighted the vulnerability of critical infrastructure to cyberattacks and the need for robust cybersecurity measures and international cooperation. As the attack spread globally, it affected numerous sectors, including shipping giant Maersk, which suffered significant operational disruptions and huge financial losses (Kaila & Nyman, 2018).

6.3.4 Cases of Cyberattack on TalkTalk and Dixons Carphone in the UK

In October 2015, TalkTalk, a major British telecommunications company, suffered a cyberattack that compromised the personal information of 157,000 customers through SQL database manipulation (Claxson, 2018). This breach exposed customer names, addresses, dates of birth, and banking information. TalkTalk was fined $400,000 by the Information Commissioner's Office (ICO) for inadequate data protection, revealing significant cybersecurity deficiencies (Boakye et al., 2023). The financial impact was substantial, costing TalkTalk an estimated $60 million in business losses, recovery costs, and reputational damage. In a separate incident, Dixons Carphone experienced a catastrophic data breach first discovered in June 2018, though it occurred in July 2017. Hackers accessed the processing systems of Currys PC World and Dixons Travel stores, compromising 5.9 million card payments and the personal information of 1.2 million customers. Although most payment cards were protected by chip-and-pin technology, approximately 105,000 non-EU cards without this protection were compromised (Claxson, 2018; Moore, 2018). Both incidents underscore the critical importance of robust cybersecurity measures and the necessity for organizations to regularly update and protect their systems against potential threats. The TalkTalk breach significantly damaged the company's reputation and eroded customer trust and loyalty. To regain trust, TalkTalk implemented strategic reputation repair tactics, including transparent communications and enhanced security measures (Boakye et al., 2023). These cases highlight the severe consequences of cyberattacks and the urgent need for proactive cybersecurity strategies and comprehensive incident response plans.

6.3.5 Cases of Cyberattack on Juspay in India

In India, Juspay is a frequently cited cybersecurity breach incident. Juspay is an Indian fintech company that processes online payments for various merchants like Amazon, Flipkart, MakeMyTrip, Uber, Airtel, and other big brands worth Rs. 1000 daily across all platforms (Deora, 2021). The breach was a significant event in the Indian fintech sector and exposed significant weaknesses in data protection practices.

The Juspay data breach in August 2020 unfortunately resulted in the exposure of over 35 million user records, including names, email addresses, and masked card data (Patil et al., 2022; Sane, Shah, & Zaveri-Shah, 2023). It is estimated that credit and debit card details of approximately one billion consumers were exposed in the incident. The leaked data included information about the user's card network (VISA/ Mastercard), the card's expiration date, the last four digits of the card, the masked card number, the card type (credit/debit card), the name on the card, and ISIN among others (Sane, Shah, & Zaveri-Shah, 2023). This incident highlighted significant vulnerabilities in India's fintech sector, where increasing digital adoption often outpaced cybersecurity measures. The breach was attributed to unauthorized access through compromised AWS server credentials (Smrithi, Doreen, & Rene, 2023). The breach led to increased scrutiny of data protection practices and required a stricter regulatory framework in India (Deora, 2021). Additionally, it highlighted the need for FinTech companies to integrate robust cybersecurity practices into their operational frameworks to prevent similar incidents. As digital transactions increase, the importance of securing sensitive user data and maintaining consumer trust cannot be overstated.

6.3.6 Cases of Cybersecurity Breaches in Africa

The Nigerian banking sector has experienced numerous cybersecurity breaches, marked by a rise in sophisticated cyberattacks such as phishing, hacking, and ATM fraud. A 2020 study revealed that viruses, worms, and Trojans were the most common breaches, driven by inadequate cybersecurity measures and regulatory compliance (Wang, Gupta, & Rao, 2015). Another study highlighted the significant financial losses and productivity impacts on Nigerian banks due to cyber fraud, underscoring the need for improved security practices and regulatory frameworks (Fatoki, 2023). South Africa, with the third highest number of cybercrime victims globally, faces substantial financial losses. The South African Banking Risk Information Center (SABRIC) reports annual losses of R2.2 billion due to cyberattacks (Olofinbiyi, 2022). A notable incident occurred in 2021, when Transaction Capital, a South African financial services company, suffered a cybersecurity breach that exposed sensitive customer information, including personal and financial data. This breach underscored the vulnerabilities in the company's cybersecurity infrastructure and highlighted the growing threat of cybercrime in South Africa's financial sector. The Transaction Capital breach prompted a comprehensive review of its security measures and calls for stricter regulatory compliance and enhanced cybersecurity protocols (Olofinbiyi, 2022). The incident also emphasized the importance of the Personal Information Protection Act (POPIA), enacted to protect personal information and ensure appropriate security measures. Following this breach, other financial institutions in South Africa reassessed and strengthened their cybersecurity frameworks, while policymakers sought to review and enhance POPIA in line with cybersecurity legislation in developed countries to bolster protection against cyber threats (Warikandwa, 2021). These cases illustrate the critical need for robust cybersecurity measures and regulatory frameworks to safeguard the financial sector in Africa from growing cyber threats.

6.3.7 Cases of Cybersecurity Breaches in Malaysia and Singapore

Malaysia has faced several significant cybersecurity breaches that have exposed vulnerabilities across various sectors as information and communications technologies (ICT) evolve. One notable incident involved the data breach reported by the Malaysia Computer Emergency Response Team (MyCERT) in 2018. This breach recorded 10,699 incidents, including fraud and attempted break-ins. Of these, fraud accounted for the highest number of incidents (5,123), while attempted break-ins were the second most reported type of incident, with 1,805 cases (Tan et al., 2020). The reported breach in Malaysia highlighted the need for robust cybersecurity measures across the country. Additionally, cybersecurity awareness and comprehensive security training of employees in the banking sector are crucial to prevent accidental disclosures and cyberattacks that result in the loss of customers and ultimately impact financial profitability (Krishnan et al., 2023). Similarly, Singapore has witnessed a surge in sophisticated cybersecurity threats despite being the most globally connected country in the world (The Business Times News, 2016; Ter, 2018). These cybersecurity breaches prompted the government to strengthen its cybersecurity framework. In 2018, Singapore enacted the Cybersecurity Act to improve the protection of its critical information infrastructure (Ter, 2018). One of the most notable breaches involved the hacking of Singapore's largest healthcare provider, SingHealth, which resulted in the theft of personal information, including that of the Prime Minister (Sia, Hosseinian-Far, & Toe, 2021).

6.3.8 Cases of Cybersecurity Breaches in Canada

Canada has faced several serious cybersecurity breaches across various sectors, highlighting the need for robust cybersecurity measures. One notable incident affected the healthcare sector, where there was a surge in cyberattacks during the COVID-19 pandemic. These attacks targeted hospitals, medical companies, and research laboratories and exploited vulnerabilities in their cybersecurity infrastructure (Wilner et al., 2021). Another major breach occurred in 2020, when cybersecurity threats were estimated to cost the world $6 trillion annually, with the number of cyberattacks increasing significantly after COVID-19. The healthcare industry has been heavily targeted due to its critical nature and the value of its data (Williams, Chaturvedi, & Chakravarthy, 2020). In addition to healthcare, Canada's information technology infrastructure has also been a focus of cybersecurity research and development. Efforts have been made to create a stable and resilient cyber infrastructure by addressing current cybersecurity challenges and developing action-oriented research topics (Craigen, Walsh, & Whyte, 2013). The banking sector in Canada has also faced cybersecurity threats, with the number of cyberattacks increasing significantly. The year 2021 was once again the worst year ever for cyberattacks in the banking sector, highlighting the urgent need for improved cybersecurity practices (Gulyas & Kiss, 2022).

6.3.9 Cases of Cybersecurity Breaches in the UAE and Gulf Countries

The United Arab Emirates (UAE) has experienced numerous cybersecurity breaches, reflecting the region's rapid digital transformation and high economic activity.

In 2018, the UAE experienced significant cyber threats, particularly targeting its oil and gas sector. The introduction of the National Institute of Standards and Technology (NIST) Cybersecurity Framework was critical to mitigating these threats and focused on key capabilities such as identifying, protecting, detecting, responding, and recovering (ALDhanhani, 2021). A major incident in 2020 affected the UAE Ministry of Defense, which thwarted a significant cyberattack on critical infrastructure. This attack highlighted the ongoing cyber threats facing the country and its proactive measures to deter and mitigate the threat through public awareness and educating internet users on cyber risks and security are integral parts of the UAE's recovery efforts and a safer cyber environment for the country (Gibbs, 2020). The UAE has introduced strict cybercrime laws and increased penalties to deter cybercriminals, demonstrating a comprehensive legislative approach to protecting its citizens and businesses (Younies & Al-Tawil, 2020). In the broader Gulf region, cyber threats are equally significant. For example, Saudi Arabia was subjected to cyberattacks on its banking sector, resulting in significant financial losses. These incidents have led to the introduction of advanced cybersecurity frameworks and international cooperation to strengthen cyber defenses against internal and external cyberattacks (Pöpper, Maniatakos, & Pietro, 2021).

6.3.10 Cases of Cybersecurity Breaches in Taiwan, China, and Japan

Taiwan has faced significant cybersecurity threats, notably from state-sponsored actors. In 2018, the APT10 Cyber Espionage Campaign linked to the Chinese government targeted Taiwan's technology and government sectors by exploiting vulnerabilities in managed IT service providers to access sensitive information (Hsini Huang & Li, 2018). The same year, Taiwan Semiconductor Manufacturing Company (TSMC) experienced a ransomware attack that temporarily halted production. In response, Taiwan implemented the Information and Communication Management Act to bolster cybersecurity defenses across public and private sectors (Hsini Huang & Li, 2018). Additionally, the Democratic Progressive Party (DPP) suffered a data breach in 2016, exposing its membership database in a politically motivated attack (Yau, 2018). By 2021, Taiwanese government websites faced approximately 5 million cyberattacks daily, including distributed denial-of-service (DDoS) attacks and attempts to breach sensitive data. Despite the high volume of attacks, Taiwan's cybersecurity infrastructure managed and mitigated these threats effectively (Jakubczak & Yau, 2021). A cyberattack on the Taipei City government systems aimed at stealing sensitive information and disrupting services prompted an upgrade of cybersecurity protocols across local government institutions (Hsini Huang & Li, 2018). Taiwan's National Cyber Security Program (2021–2024) aims to enhance cybersecurity standards, focusing on critical infrastructure protection and international cooperation (Jakubczak & Yau, 2021). The government adopts a centralized approach to cybersecurity with comprehensive compliance requirements for all government agencies and critical infrastructure providers (Hsini Huang & Li, 2018). Taiwan actively participates in international cybersecurity collaborations, including joint cyber exercises with the United States, and invests in cybersecurity education and workforce development to cultivate a skilled cybersecurity workforce. Public-private partnerships are emphasized to enhance overall security resilience (Ku, Chang, & Yen, 2009; Jakubczak & Yau, 2021).

China is a major player in the global cybersecurity landscape and is often accused of cyber espionage targeting foreign governments and companies. A significant case was the breach of the Office of Personnel Management (OPM) in the United States, attributed to Chinese hackers, which compromised the sensitive information of millions of US federal employees (Lindsay, 2015). Another notable incident was the Microsoft Exchange server hack in 2021, affecting thousands of organizations worldwide and demonstrating China's cyber capabilities (Nasheri, 2023). China's cybersecurity law, enacted in 2016, aims to strengthen cyberspace governance through privacy and security measures. However, it has been seen as a barrier to foreign companies due to requirements for data localization, cooperation with law enforcement, and content restrictions (Qi, Shao, & Zheng, 2018). The Chinese government's emphasis on cyber sovereignty and cyberspace regulation highlights its broader strategy to achieve technological supremacy and national security (Cai, 2015; Ahmed & Weber, 2018).

Japan has also been a key target for cyberattacks due to its advanced technological infrastructure and geopolitical importance. The 2015 cyberattack on the Japan Pension Service compromised the personal information of millions, underscoring the vulnerability of Japanese institutions (Nitta, 2021). Japan's cybersecurity policy has evolved to address these challenges, focusing on defensive measures and international cooperation. The Japanese government has proactively developed its cybersecurity infrastructure and policies, becoming a significant player in global cybersecurity (Aikawa, 2020).

6.4 EFFECT OF CYBERSECURITY BREACHES AND INSIDER THREATS ON ENTREPRENEURSHIP

The increasing use of digital technologies by entrepreneurs globally has led to significant cybersecurity breaches and insider threats, posing major challenges to entrepreneurship. These breaches can result in business interruptions, loss of customers, financial losses, reputational damage, and even bankruptcy, particularly affecting SMEs due to their limited resources (Raimi, 2022). In developed countries, the financial impact of cyberattacks is substantial, with infrastructure outages costing an average of $100,000 per hour and critical application outages costing between $500,000 and $1 million per hour (IBM Services, 2020). The United States Federal Bureau of Investigation (FBI) reported over 4,000 daily ransomware attacks, while cybercriminals create more than 330,000 dangerous malware applications daily (Hasan et al., 2021). SMEs often face ransom demands averaging around $116,000 (Witts, 2022).

Cybersecurity breaches also significantly impact business innovation, as companies often reduce research and development spending following a breach, hindering long-term growth. In developing countries, the consequences are even more severe due to weaker institutional frameworks and lower cybersecurity preparedness. SMEs in these regions frequently lack adequate cybersecurity measures, making them highly vulnerable to cyberattacks. The limited perception of cybersecurity importance among these companies exacerbates the situation, leading to frequent breaches and disruptions in business operations (Kabanda, Tanner, & Kent, 2018). Overall, the growing digital landscape exposes entrepreneurs to heightened cybersecurity risks, necessitating improved cybersecurity measures, especially for SMEs. Addressing these vulnerabilities is crucial to protecting business continuity, fostering innovation, and ensuring long-term growth in both developed and developing regions. Another negative impact of cybersecurity breaches is the disruption of supply chains, leading to operational, economic, and transactional risks (Melnyk et al., 2021). Likewise, insider threats, where employees or partners gain access to a company's data and systems, pose a significant cybersecurity challenge. Insider threats have resulted in significant financial harm to financial institutions (Wang, Gupta, & Rao, 2015; Efijemue, Ejimofor, & Owolabi, 2023).

In developing countries, insider threats are exacerbated by a lack of cybersecurity awareness and employee training. For example, a study in South Africa found that personal norms and attitudes significantly influence the likelihood that individuals will commit cybersecurity misconduct, highlighting the need for improved employee training and awareness programs (Silaule, Makhubele, & Mamorobela, 2022). From a macroeconomic perspective, the economic impact of cybersecurity breaches and insider threats is significant, affecting both current business operations and future growth prospects. In developed countries, integrating cybersecurity into national economic policies is critical. Cybersecurity breaches disrupt critical infrastructure and the functioning of the markets (Timcke, Gaffley, & Rens, 2023). The weak institutional environment for cybersecurity is known to hinder entrepreneurial development, as entrepreneurs in such regions often struggle with inadequate support systems to protect against cyber threats, hindering their ability to innovate, compete, and grow. Figure 6.2 illustrates the impact of cybersecurity breaches and insider threats on entrepreneurship.

Figure 6.2 highlights the impact of cybersecurity breaches and insider threats on entrepreneurship, affecting both developed and developing countries. In developing nations, weak cybersecurity infrastructure, lack of employee training, and the vulnerability of SMEs lead to frequent breaches. Meanwhile, developed countries face significant financial losses, reputational damage, and high ransom demands. Insider threats, including employee misconduct and partner issues, further compromise security by bypassing protective measures. The consequences of breaches include disrupted supply chains, loss of customers, business interruptions, and a decline in research and development spending.

6.5 STRATEGIES FOR CYBERSECURITY THREATS AND RESILIENCE

The recurring threats to the cyber landscape require robust strategies to ensure cybersecurity resilience. These strategies include information sharing and collaboration, cybersecurity education and training, international collaboration, and cybersecurity insurance for business recovery. Figure 6.3 shows cybersecurity threat and resiliency strategies.

Figure 6.3 shows strategies for addressing cybersecurity threats and building resilience. It highlights key approaches such as education and training, including ongoing training to mitigate human error and a proactive resilience framework. International

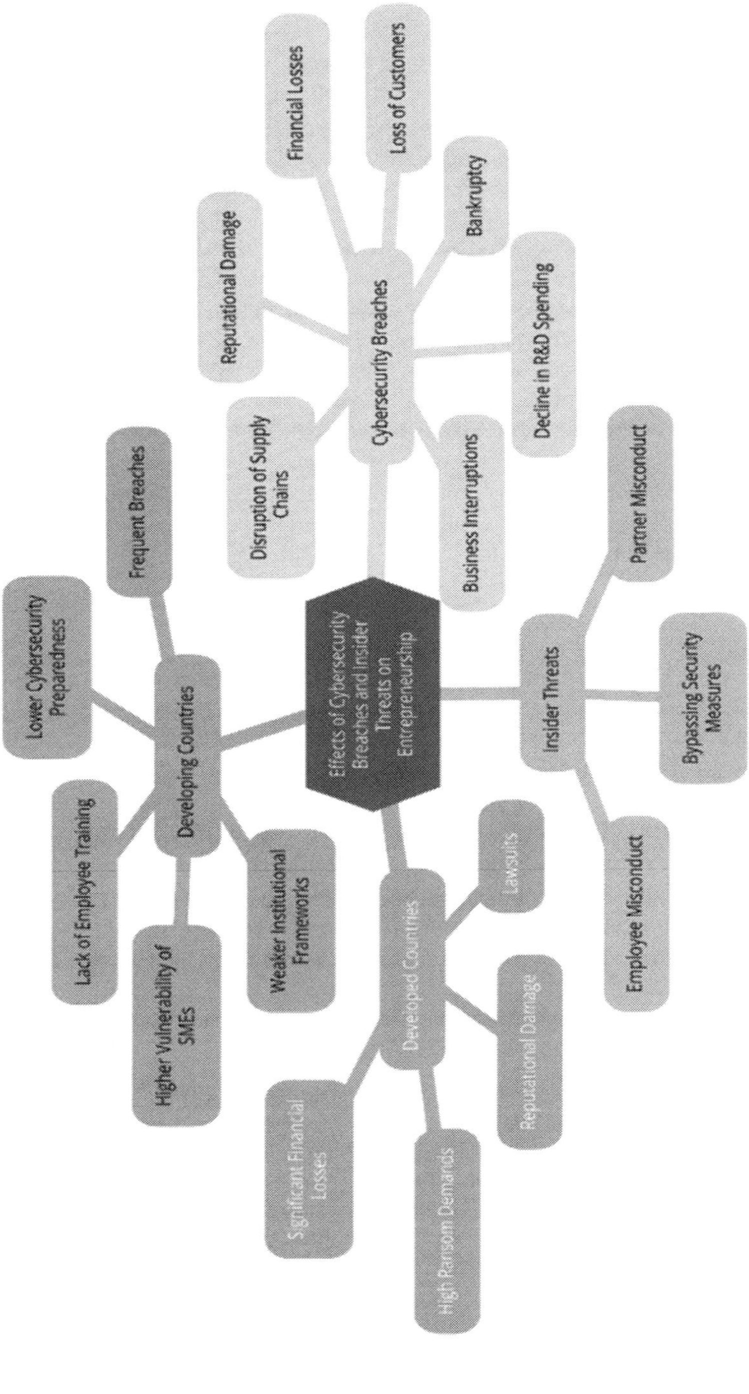

FIGURE 6.2 Cybersecurity breaches, insider threats, and entrepreneurship.

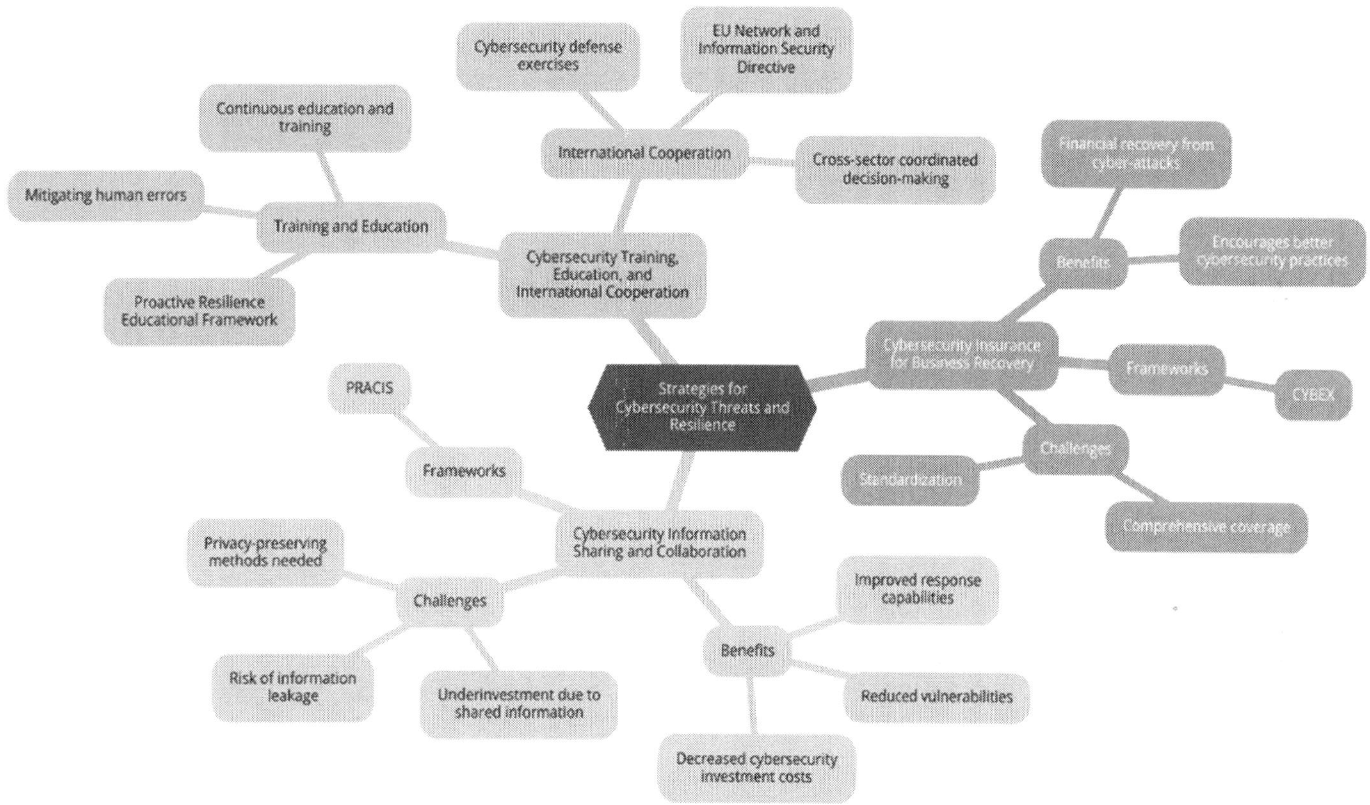

FIGURE 6.3 Cybersecurity threat and resiliency strategies by author.

cooperation and cross-sectoral decision-making, supported by the EU Network and Information Security Directive, are crucial for defense coordination. Sharing and collaborating on cybersecurity information helps reduce costs and vulnerabilities, although privacy concerns and under-investment present challenges. The illustration also addresses cybersecurity insurance for business recovery, promoting better practices and financial recovery from cyberattacks.

6.5.1 Cybersecurity Information Sharing and Collaboration

Sharing information between organizations is a critical strategy for improving cybersecurity resilience. Effective collaboration can significantly reduce vulnerabilities and improve response capabilities. Studies have shown that organizations that are willing to share information are more likely to collaborate effectively to address cybersecurity threats. Furthermore, willingness to share information correlates with increased integrative efforts across organizations, highlighting the importance of fostering a culture of information sharing within organizations. A game theory approach to information sharing, particularly in cloud computing environments, highlights the benefits of collaborative cybersecurity efforts. Sharing cyber threat intelligence can reduce the likelihood of multiple attacks exploiting the same vulnerability in different organizations, thereby reducing overall cybersecurity investment costs. However, challenges such as the risk of information leakage and the possibility that some companies may underinvest in cybersecurity due to reliance on shared information need to be addressed. Additionally, privacy-preserving methods for information sharing are essential to encourage participation while protecting sensitive data. The PRACIS framework, which integrates privacy-preserving and homomorphic encryption techniques, demonstrates how secure information sharing can be achieved to ensure organizations provide valuable threat intelligence without compromising privacy. Similar efforts to secure information sharing are critical to maintaining trust and promoting collaboration between organizations.

6.5.2 Cybersecurity Training, Education, and International Cooperation

Cybersecurity education and training are important parts of building resilience. A comprehensive educational framework such as the Proactive Resilience Educational Framework (Prosilience EF) can improve the knowledge and skills of healthcare IT personnel with an emphasis on interactive and contextual training. This framework highlights the importance of continuous education and training to mitigate human errors and improve the overall cybersecurity posture. International cooperation also plays an important role in cybersecurity resilience. Cross-border collaboration can lead to better preparedness and response to cyber threats. For example, the EU Network and Information Security Directive requires cybersecurity cooperation between member states, improving situational awareness and incident management. Furthermore, coordinated decision-making between different sectors, including intelligence, law enforcement, and public health, is critical to addressing complex cyber threats. Cybersecurity defense exercises focused on improving threat information sharing can help develop critical soft skills in cybersecurity specialists. Addressing educational gaps and fostering a culture of collaboration can significantly increase the effectiveness of these exercises.

6.5.3 Cybersecurity Insurance for Business Recovery

Cybersecurity insurance is a new strategy to mitigate the financial impact of cyber incidents and support business recovery. Cyber insurance policies can cover losses from cyberattacks and help companies recover faster. Integrating cyber insurance into information exchange frameworks such as Cybersecurity Information Exchange (CYBEX) can improve overall risk management. By using game theory models, CYBEX can incentivize participation and optimize threat intelligence sharing, ultimately leading to better cybersecurity outcomes. The benefits of cybersecurity insurance go beyond financial recovery. Additionally, organizations are encouraged to adopt better cybersecurity practices to qualify for insurance coverage, thereby improving their resilience to cyber threats. However, challenges such as standardization and ensuring comprehensive coverage remain areas for further research and development.

6.6 CONCLUSION

The conclusion of this chapter addresses the impact of cybersecurity breaches and insider threats on knowledge management in entrepreneurial ecosystems. In various global contexts, cybersecurity incidents highlight the urgent need for robust knowledge management practices in the following respects. First, analysis of significant breaches such as the Equifax breach in the US, the NotPetya attack in Ukraine, and the TalkTalk breach in the UK show that the dissemination and integration of cybersecurity knowledge is of paramount importance. These cases highlight the importance of fostering a cybersecurity culture, implementing strong regulatory frameworks, and promoting international cooperation to share threat intelligence. For entrepreneurs, particularly in developing countries, the challenge is compounded by limited resources and technical skills. Incidents such as the Nigerian banking sector breaches and the Juspay breach in India highlight the need for strong regulations and a focus on integrating cybersecurity knowledge into business operations. Likewise, the transaction capital breach in South Africa and the breaches in Israel and Singapore underscore the need for continued learning and collaboration in the face of evolving threats. Effective knowledge management practices are essential to mitigating the risks associated with these cybersecurity threats. Since entrepreneurs often lack sufficient cybersecurity knowledge, which makes their businesses particularly vulnerable, deliberate decisions by policymakers and managers to integrate cybersecurity training into entrepreneurship education can significantly improve the resilience of new businesses. Additionally, the use of advanced technologies such as artificial intelligence and machine learning can improve insider threat detection and response, creating a more secure operating framework. Furthermore, developing a strong cybersecurity culture within organizations is critical. This includes raising awareness and training employees to prevent insider threats and ensure the secure handling of information. Collaboration and information sharing between companies and across borders are also critical to building a resilient cybersecurity infrastructure. In summary, addressing the impact of cybersecurity breaches and insider threats requires a multi-pronged approach that integrates sound knowledge management practices, ongoing education and international collaboration. These strategies are essential to protecting business ecosystems from the ever-evolving cyber threat landscape.

REFERENCES

Admass, W. S., Munaye, Y. Y., Diro, A. A. 2024. Cyber security: State of the art, challenges and future directions, *Cyber Security and Applications*, *2*. https://doi.org/10.1016/j.csa.2023.100031

Ahmed, S., & Weber, S. (2018). China's long game in techno-nationalism. *First Monday*. https://firstmonday.org/ojs/index.php/fm/article/download/8085/7209

Aikawa, W. (2020). Japan's cybersecurity policy. *Telecommunications Policies of Japan*, *20* 133–148.

ALDhanhani, M. J. (2021). Review of cyber security on oil and gas industry in United Arab Emirates: Analysis on the effectiveness of the national institute of standards and technology's (NIST) cybersecurity framework. *Turkish Journal of Computer and Mathematics Education*, *12*(11), 714–720.

Alnajim, A. M., Habib, S., Islam, M., Thwin, S. M., & Alotaibi, F. (2023). A comprehensive survey of cybersecurity threats, attacks, and effective countermeasures in industrial Internet of Things. *Technologies*, *11*(6), 161. https://doi.org/10.3390/technologies11060161

Barbas, J. M. A. 2024. Knowledge management and continuous improvement in cyberspace. In F. A. Carneiro Pacheco de Andrade, P. M. Fernandes Freitas, & J.R. de Sousa Covelo de Abreu (eds) *Legal developments on cybersecurity and related fields*. Law, Governance and Technology Series, vol 60. Springer, Cham. https://doi.org/10.1007/978-3-031-41820-4_5

Bates, A., & Hassan, W. U. (2019). Can data provenance put an end to the data breach? *IEEE Security & Privacy*, *17*(6), 88–93. https://doi.org/10.1109/MSEC.2019.2913693

Batmetan, J. R. & Kembuan, D. R. E. (2024). Effects of knowledge-sharing methods on cyber security practice in small medium enterprises. *Proceedings of the 5th Vocational Education International Conference (VEIC-5 2023)*, Advances in Social Science, Education and Humanities Research, Atlantis Press – part of Springer Nature. https://doi.org/10.2991/978-2-38476-198-2_5. https://www.atlantis-press.com/proceedings/veic-23/125997779

Boakye, D., Sarpong, D., Meissner, D., & Ofosu, G. (2023). How TalkTalk did the walk-walk: strategic reputational repair in a cyber-attack. *Information Technology & People*. https://doi.org/10.1108/ITP-08-2022-0589

Bryda, G. (2020). Whats and hows? The practice-based typology of narrative analyses. *Przegląd Socjologii Jakościowej*, *16*(3), 120–142.

Cai, C. (2015). Cybersecurity in the chinese context: Changing concepts, vital interests, and prospects for cooperation. *International Journal of Cyber Security*, *1*, 471–496.

Catota, F. E., Morgan, M. G., & Sicker, D. (2018). Cybersecurity incident response capabilities in the Ecuadorian financial sector, *Journal of Cybersecurity*, *4*(1), tyy002. https://doi.org/10.1093/cybsec/tyy002

Claxson, N. (2018). Securing VoIP: Encrypting today's digital telephony systems. *Network Security*, *2018*(3), 11–13.

Craigen, D., Walsh, J. D., & Whyte, D. (2013). Securing Canada's information-technology infrastructure: Context, principles, and focus areas of cybersecurity research. *Technology Innovation Management Review*, *3*(1), 12–18.

CSI Computer Security Institute. (2011). *CSI computer crime and security survey 2010/2011*. Computer Security Institute.

Dasawat, S. S., & Sharma, S. (2023). Cyber security integration with smart new age sustainable startup business, risk management, automation and scaling system for entrepreneurs: An artificial intelligence approach. *2023 7th International Conference on Intelligent Computing and Control Systems (ICICCS)*, 1357–1363.

Deora, K. (2021). India's need for ethical vulnerability research. *International Research Journal of Engineering and Technology (IRJET)*, *8*(7), 1443–1446

Efijemue, O., Ejimofor, I., & Owolabi, O. S. (2023). Insider threat prevention in the US banking system. *International Journal on Soft Computing*, *14*(3), 17–29.

Falowo, O. I., Popoola, S., Riep, J. R., Adewopo, V., & Koch, J. (2022). Threat actors' tenacity to disrupt: examination of major cybersecurity incidents. *IEEE Access*, *10*, 134038–134051.

Fatoki, J. O. (2023). The influence of cyber security on financial fraud in the Nigerian banking industry. *International Journal of Science and Research Archive*, *9*(2), 503–515.

Ferland, J. (2019). Cyber insurance–What coverage in case of an alleged act of War? Questions raised by the Mondelez v. Zurich case. *Computer Law & Security Review*, *35*(4), 369–376.

Gibbs, T. (2020). Seeking economic cyber security: A Middle Eastern example. *Journal of Money Laundering Control*, *23*(3), 493–507.

Granato, A., & Polacek, A. (2019). The growth and challenges of cyber insurance. *Chicago Fed Letter*, *426*, 1–6. https://doi.org/10.21033/cfl-2019-426

Greitzer, F. L., Moore, A. P., Cappelli, D. M., Andrews, D. H., Carroll, L. A., & Hull, T. D. (2008). Combating the insider cyber threat. *IEEE Security & Privacy*, *6*(1), 61–64. https://doi.org/10.1109/MSP.2008.8

Gulyas, O., & Kiss, G. (2022). Cybersecurity threats in the banking sector. *2022 8th International Conference on Control, Decision and Information Technologies (CoDIT)*, 1, 1070–1075.

Hasan, S., Ali, M., Kurnia, S., & Thurasamy, R. (2021). Evaluating the cyber security readiness of organisations and its influence on performance. *Journal of Information Security and Applications*, *58*, 102726.

Hecker, J. & Kalpokas, N. (2024). Thematic analysis vs. narrative analysis. online guide, ATLAS.ti Scientific Software Development GmbH. https://atlasti.com/guides/thematic-analysis/thematic-analysis-vs-narrative-analysis

Hsini Huang, & Li, T. (2018). A centralised cybersecurity strategy for Taiwan. *Journal of Cyber Policy*, *3*(3), 344–362.

Hüsch, P. & Sullivan, J. (2023). Global approaches to cyber policy, legislation and regulation a comparative overview, *Royal United Services Institute for Defence and Security Studies*, April 2023. https://static.rusi.org/rusi-global-approaches-to-cyber-special-resource_0.pdf

Huzaizi, A. H. A., Tajuddin, S. N. A. A., Bahari, K. A., Manan, K. A., & Mubin, N. N. A. (2021). Cyber-security culture towards digital marketing communications among small and medium-sized (SME) entrepreneurs. *Asian Culture and History*, *13*(2), 20–28.

IBM Services (2020) Adapt and respond to risks with a business continuity plan (BCP). https://www.ibm.com/au-en/services/business-continuity/plan

Jakubczak, W., & Yau, H. M. (2021). Trends in cybersecurity regulations of Taiwan (Republic of China)–Phases of promotion of major cyber security plans and programs in the national cyber security program of Taiwan (2021–2024). Zeszyty Naukowe SGSP/Szkoła Główna Służby Pożarniczej.

Kabanda, S., Tanner, M., & Kent, C. (2018). Exploring SME cybersecurity practices in developing countries. *Journal of Organizational Computing and Electronic Commerce*, *28*(3), 269–282.

Kaila, U., & Nyman, L. (2018). Information security best practices: First steps for startups and SMEs. *Technology innovation management review*, *8*(11), 32–42.

Käss, S., Brosig, C., Westner, M. et al. (2024). Short and sweet: Multiple mini case studies as a form of rigorous case study research. *Information Systems e-Business Management, X*(Y). https://doi.org/10.1007/s10257-024-00674-2

Kenny, C. (2018). The equifax data breach and the resulting legal recourse. *The Brooklyn Journal of Corporate, Financial and Commercial Law*, 13(1), 216–238.

Kostyuk, N. (2024). Allies and diffusion of state military cybercapacity. *Journal of Peace Research*, 61(1), 44–58. https://doi.org/10.1177/00223433241226559

Krishnan, S. G., Al-Nahari, A., Ismail, N. A., & Yao, D. N. L. (2023). Enhancing cybersecurity awareness among banking employees in Malaysia: Strategies, implications, and research insights. *International Journal of Academic Research in Business and Social Sciences, 13*(8), 596–612.

Ku, C., Chang, Y., & Yen, D. (2009). National information security policy and its implementation: A case study in Taiwan. *Telecommunications Policy, 33*(371–384), 291–303.

Kumar, N., Sen, A., Hordiichuk, V., Jaramillo, M., Molodetskyi, B., & Kasture, A. (2023). AI in cybersecurity: Threat detection and response with machine learning. *Tuijin Jishu/Journal of Propulsion Technology, 44*(3), 38–46.

Lika, R. A., Murugiah, D., Brohi, S., & Ramasamy, D. (2018). NotPetya: Cyber attack prevention through awareness via gamification. *2018 International Conference on Smart Computing and Electronic Enterprise (ICSCEE)*, 1–6.

Lindsay, J. (2015). The impact of China on cybersecurity: Fiction and friction. *International Security, 39*, 7–47.

Luszcz, J. (2018). Apache struts 2: How technical and development gaps caused the equifax breach. *Network Security, 2018*(1), 5–8.

Maesschalck, S. (2024). Gentlemen, you can't fight in here. Or can you?: How cyberspace operations impact international security. *World Affairs, 187*(1), 24–36. https://doi.org/10.1002/waf2.12004

Melnyk, S., Schoenherr, T., Speier, C., Peters, C., Chang, J. F., & Friday, D. (2021). New challenges in supply chain management: Cybersecurity across the supply chain. *International Journal of Production Research, 60*, 162–183.

Moore, M. (2018). *Dixons Carphone hit with major data breach*. TechRadar. Retrieved from TechRadar. https://www.techradar.com/news/dixons-carphone-hit-with-major-data-breach

Nasheri, H. (2023). State-sponsored economic espionage in cyberspace: Risks and preparedness. In *Cybercrime in the pandemic digital age and beyond*, Russell G. Smith, Rick Sarre, Lennon Yao-Chung Chang, Laurie Yiu-Chung Lau (eds.) (pp. 87–107). Springer International Publishing.

Nicho, M., & Kamoun, F. (2014). Multiple case study approach to identify aggravating variables of insider threats in information systems. *Communications of the Association for Information Systems*, 35(1), 18, 1–123.

Nie, D., & Xu, C. (2021). Non-GAAP earnings quality in firms with data breach incident. *Asian Review of Accounting*, 29(3), 383–398.https://doi.org/10.1108/ara-10-2020-0169

Nitta, Y. (2021). The evolving concept of the Japanese security strategy. In Paul Cornish (ed.), *The Oxford handbook of cyber security, Oxford handbooks* (2021; online edn, Oxford Academic, 8 Dec. 2021), https://doi.org/10.1093/oxfordhb/9780198800682.013.42, accessed 11 June 2024.

Niu, B., Wang, L., Yu, X., Feng, B. 2024. Data-driven analysis of digital entrepreneurship in medical supply resilience confronting the COVID-19 epidemic, *Information Processing & Management*, 61(1), 2024. https://doi.org/10.1016/j.ipm.2023.103502

Novak, A., & Vilceanu, O. (2019). "The internet is not pleased": twitter and the 2017 Equifax data breach. *The Communication Review*, 22(3), 196–221

Olofinbiyi, S. A. (2022). A reassessment of public awareness and legislative framework on cybersecurity in South Africa. *ScienceRise: Juridical Science*, 2(20), 34–42.

Patil, K., Sonune, H., Devikar, S., Chaudhari, V., & Ayachit, I. (2022). A comparative analysis of various techniques of data leakage detection in different domains. In ICT *analysis and applications: Proceedings of ICT4SD 2022* (pp. 735–742). Springer Nature Singapore.

Pöpper, C., Maniatakos, M., & Pietro, R. D. (2021). Cyber security research in the Arab region. *Communications of the ACM, 64*(5), 96–101.

Prabhu, S., & Thompson, N. (2022). A primer on insider threats in cybersecurity. *Information Security Journal: A Global Perspective, 31*(5), 602–611.

Qi, A., Shao, G., & Zheng, W. (2018). Assessing China's cybersecurity law. *Computer Law & Security Review, 34*, 1342–1354.

Raimi, L. (2022) Business continuity and disaster recovery strategies (BCRS) as resilience tools after cyberattacks in entrepreneurship ecosystems. How ready is peaceful East Asia? Proceeding of RIN Online Workshop Proceedings held on August 31, 2022. https://d-arch.ide.go.jp/RIN/common/pdf/2022_rin-ws-proceedings.pdf

Raimi, L. (2023). Business continuity and disaster recovery strategies as resilience tools after cyberattacks in toxic enterpreneurship ecosystems. In *Cybersecurity for decision makers* (pp. 349–364). CRC Press.

Raineri, E. M., & Fudge, T. P. (2019). Exploring the sufficiency of undergraduate students' cybersecurity knowledge within top universities' entrepreneurship programs, *Journal of Higher Education Theory and Practice, 8*, 176–188.

Riessman, C. K. (2020). Looking back, looking forward. *Narrative Works, 10*, 117–124.

Sane, R., Shah, A., & Zaveri-Shah, B. (2023). Should consumers be prohibited from storing card data on the internet? *Indian Law Review, 7*(3), 281–298.

Sharifi, S. (2023). A novel approach to the behavioral aspects of cybersecurity. arXiv preprint arXiv:2303.13621. https://arxiv.org/pdf/2303.13621

Sia, N. C., Hosseinian-Far, A., & Toe, T. T. (2021, May). Reasons behind poor cybersecurity readiness of Singapore's small organizations: Reveal by case studies. In *Cybersecurity, Privacy and Freedom Protection in the Connected World: Proceedings of the 13th International Conference on Global Security, Safety and Sustainability*, London, January 2021 (pp. 269–283). Springer International Publishing.

Silaule, C. B., Makhubele, L. M., & Mamorobela, S. (2022). A model to reduce insider cybersecurity threats in a South African telecommunications company. *SA Journal of Information Management, 24*(1), 1–12. https://doi.org/10.4102/sajim.v24i1.1573

Sivrieva, G. (2018). The equifax breach amid a lawless landscape: Changes are afoot for privacy & data security due to the European Union's general data protection regulation. *Wayne Law Review*, 64, 553.

Smrithi, S., Doreen, H. M. D., & Rene, R. C. R. (2023). An analysis of the increasing cases of data breaches in India. *i-manager's Journal on Software Engineering, 34*, 312–319.

Stamp, J., & Mwapwele, S.D. (2024). Examining data governance to determine how democratic data management can be achieved in organizations. In A. Gerber (Eds.), *South African computer science and information systems research trends. SAICSIT 2024. Communications in Computer and Information Science* (Vol. 2159). Springer, Cham. https://doi.org/10.1007/978-3-031-64881-6_25

Tan, O. S. L., Vergara, R. G., Phan, R. C., Khan, S., & Khan, N. (2020). Cybersecurity laws in Malaysia. In Encyclopedia of criminal activities and the deep web (pp. 435–448). IGI Global.

Ter, K. L. (2018). Singapore's cybersecurity strategy. *Computers & Security, 34*(5), 924–927.

The Business Times News (2016, February) McKinsey report ranks Singapore as world's most connected country. https://www.businesstimes.com.sg/startups-tech/technology/mckinsey-report-ranks-singapore-worlds-most-connected-country

Timcke, S., Gaffley, M., & Rens, A. (2023). The centrality of cybersecurity to socioeconomic development policy: A case study of cyber-vulnerability at South Africa's Transnet. *The African Journal of Information and Communication (AJIC), 32*, 50–72.

Wang, J., Gupta, M., & Rao, H. R. (2015). Insider threats in a financial institution. *MIS quarterly*, *39*(1), 91–112.
Wang, S., & Wang, H. (2019). Knowledge management for cybersecurity in business organizations: A case study. *Journal of Computer Information Systems*, *61*(2), 174–181.
Warikandwa, T. (2021). Personal data security in South Africa's financial services market: The protection of personal information act 4 of 2013 and the European Union general data protection regulation compared. *Potchefstroom Electronic Law Journal*, 24, 1–32.
Wibisono, E. (2023). The digital entrepreneurial ecosystem in the European Union: evidence from the digital platform economy index. *European Planning Studies*, 31(6), 1270–1292. https://doi.org/10.1080/09654313.2023.2202683
Williams, C. M., Chaturvedi, R., & Chakravarthy, K. (2020). Cybersecurity risks in a pandemic. *Journal of Medical Internet Research*, *22*(1), s. doi:10.2196/23692
Wilner, A. S., Luce, H., Ouellet, E., Williams, O., & Costa, N. (2021). From public health to cyber hygiene: Cybersecurity and Canada's healthcare sector. *International Journal*, *76*(4), 522–543.
Witts, J. (2022) The top 5 biggest cyber security threats that small businesses face and how to stop them. *Expert Insights Article*. https://expertinsights.com/insights/the-top-5-biggest-cyber-security-threats-that-small-businesses-face-and-how-to-stop-them/
Yau, H. M. (2018). Explaining Taiwan's cybersecurity policy prior to 2016: Effects of norms and identities. *Issues & Studies*, 54(02), 1–30. https://doi.org/10.1142/S1013251118500042
Younies, H., & Al-Tawil, T. (2020). Effect of cybercrime laws on protecting citizens and businesses in the United Arab Emirates (UAE). *Journal of Financial Crime*, *27*(4), 1089–1105.

7 The Language of Hate
An Exploration of Cyberbullying and Hate Speech Using Artificial Intelligence Techniques

Swati Maste, Chinmayi Shetty, Pallavi Prabhu, Richa Sharma, and Arti Arya

7.1 INTRODUCTION

Social networks have revolutionized communication, enabling people worldwide to connect, share ideas, and engage with diverse perspectives. Unfortunately, this digital landscape has also given rise to a dark underbelly: the pervasive presence of hate speech and cyberbullying. These destructive phenomena pose significant threats to individuals and communities, warranting urgent attention and intervention. The impact of hate speech cannot be understated. Studies reveal that online hate speech incidents have surged by 20% since the start of the pandemic (BBC News, 2021). The virtual world provides a breeding ground for individuals to unleash their vitriol, targeting individuals based on their ethnicity, race, gender, religion, or other defining characteristics. This amplification of hate speech online not only perpetuates discrimination but also fuels a culture of intolerance and division.

In parallel, cyberbullying has become an all-too-common experience, particularly among younger generations. Astonishingly, research indicates that a staggering 73% of students have experienced cyberbullying, with a significant portion enduring such abuse on multiple occasions. Statistics reveal that a substantial 37% of teenagers have personally encountered cyberbullying, prompting significant concerns (Patchin, 2019). The consequences extend beyond immediate emotional distress, as cyberbullying victims are at higher risk of developing anxiety, depression, and even suicidal tendencies (Broadband Search, 2023). In this chapter, we embark on an in-depth exploration of cyberbullying and hate speech, dissecting their impact on individuals, communities, and society. We delve into the psychological, social, and legal ramifications of these phenomena, uncovering the intricate dynamics at play. Additionally, we examine detection techniques, leveraging advances in machine learning (ML), natural language processing, and artificial intelligence, to detect and address instances of cyberbullying and hate speech.

DOI: 10.1201/9781003498094-7

Through a comprehensive understanding of these issues and the implementation of effective countermeasures, we can forge a path toward a more harmonious online world. By shedding light on the language of hate, we aim to empower individuals, institutions, and policymakers to confront and eradicate cyberbullying and hate speech, fostering digital spaces that promote empathy, understanding, and mutual respect.

7.2 UNDERSTANDING HATE SPEECH AND CYBERBULLYING

7.2.1 Hate Speech

Hate speech refers to any kind of speech, action, or gesture that threatens, denigrates, or insults a person or group based on their religion, race, ethnicity, gender, sexual orientation, or handicap. It is a specific sort of insulting language where the speaker grounds his viewpoints based on these preconceived notions, or stereotypes. The existence of diverse types of hate speech, such as racial, religious, gender-based, and many others, results in distinct consequences for individuals as well as for society. For instance, racial hate speech can contribute to feelings of marginalization and perpetuate systemic discrimination, while religious hate speech may foster intolerance and hinder freedom of worship. Understanding these variations is crucial for developing targeted interventions and promoting a more inclusive and harmonious society. Some of the most common types of Online Hate Speech are:

A. Racial or Ethnic Hate Speech: Involves singling out individuals or a community based on their race or ethnicity and asserting the superiority of one race over others. It perpetuates social divisions and undermines the principles of equality and human rights.
B. Religious Hate Speech: Refers to the act of insulting or inciting hatred toward individuals or groups based on their religious beliefs or practices. This form of hate speech can fuel religious intolerance, contribute to interfaith tensions, and undermine freedom of religion and belief, ultimately hindering social harmony and cultural diversity.
C. Homophobic/Transphobic/Gender-based Hate Speech: Involves using depreciative language, slurs, or discriminatory remarks against individuals based on their gender or sexual orientation. According to a report, approximately 64% of social media users identifying as LGBTQA+ have reported instances of harassment and hate speech, a significantly higher rate compared to all other identity groups (NPR, 2021).
D. Disability-based Hate Speech: It is any form of statement that expresses hatred or hostility toward people with disabilities. This can include verbal abuse, written insults, threats, or any other form of communication that is intended to harm or intimidate people with disabilities.

Addressing all these and many other types of hate speech is crucial for promoting inclusivity, respect, equal rights, and peace for all individuals, regardless.

7.2.2 Cyberbullying

Cyberbullying is a form of online harassment where an individual purposefully causes harm to another person through various digital platforms. It typically involves sending threatening messages, spreading rumors, sharing private or embarrassing information, or engaging in social exclusion.

The effects of hate speech and cyberbullying can be far-reaching. They may negatively affect the victims' mental health., as well as the social fabric of society. It may make it difficult for people to feel secure and can foster an atmosphere of fear and intimidation. Here are a few possible effects it might have on people:

A. Emotional and Psychological Impact: Cyberbullying can lead to severe emotional distress, anxiety, and depression in the victims. It may also result in feelings of isolation, fear, and helplessness, negatively impacting their overall mental well-being.
B. Social Isolation and Relationship Strain: Victims of cyberbullying often feel socially isolated as they may be subjected to online ostracism or be shunned by their peers. The fear of being targeted or humiliated can cause them to withdraw from social activities and relationships, leading to a strained social life and a sense of loneliness.
C. Self-harm and Suicidal Tendencies: In extreme cases, cyberbullying can push victims to the brink of self-harm or suicide. The relentless and public nature of online abuse can make victims feel hopeless and trapped. It is crucial to take any indications of self-harm or suicidal ideation seriously and seek immediate help.

Addressing the effects of cyberbullying requires a multi-faceted approach involving education, awareness, and support systems. By creating a safe and inclusive online environment, we can attenuate the devastating consequences of cyberbullying and protect the well-being of individuals in the digital age. Figure 7.1 illustrates the progression of cyberbullying and hate speech from the year 2004–2022, revealing a concerning escalation in recent years. The data indicates a significant increase in cyberbullying from 2004 to 2013, which remained at elevated levels until reaching its highest point in 2022. In contrast, hate speech experienced a gradual rise over the years, with a notable escalation after 2013, peaking in 2016 and continuing thereafter. The data underscores a deteriorating trend, emphasizing the urgent need for mitigation measures to be implemented promptly.

7.3 IMPACT OF CYBERBULLYING AND HATE SPEECH ON MENTAL HEALTH

Exposure to online hate speech and cyberbullying can have profound psychological effects on individuals, contributing to mental health issues such as suicidal ideation, depression, and anxiety. Research has shown that individuals who experience online harassment and bullying are at a higher risk of contemplating self-harm or engaging in suicidal behaviors (Hamm et al., 2015). Based on statistical

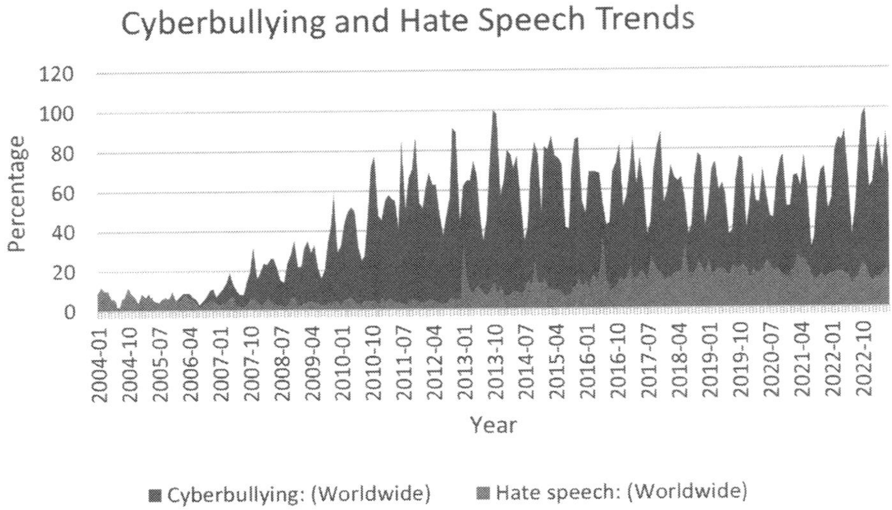

FIGURE 7.1 Examining the escalation of cyberbullying and hate speech (2004–2022).

data, 41% of children attribute their social anxiety to the effects of cyberbullying. The relentless nature of online attacks and the anonymity of perpetrators can intensify feelings of hopelessness, social isolation, and despair, leading to an increased vulnerability to suicidal thoughts.

Depression is a common consequence of online hate speech and cyberbullying. Victims may suffer ongoing melancholy, a loss of interest in activities, alterations in eating and sleep patterns, and attention problems (Hinduja & Patchin, 2017). Exposure to negative and derogatory online content can erode self-esteem, foster negative self-perception, and exacerbate symptoms of depression. Anxiety is also prevalent among those targeted by online aggression. The constant threat of being targeted or humiliated online can create a pervasive sense of anxiety, impacting daily functioning and overall well-being. Figure 7.2 describes the detrimental effects of cyberbullying and hate speech propagation on social media and online platforms.

Addressing the clinical and psychological aspects of mental health resulting from online hate speech and cyberbullying is crucial. Building awareness, promoting empathy, and fostering a culture of digital respect are essential in mitigating the negative impact on mental health. Furthermore, creating accessible mental health resources and promoting education about the psychological consequences of online aggression can help individuals cope with the emotional toll and seek appropriate support.

7.3.1 Legal and Ethical Considerations in Hate Speech Regulation

The extent of online hate speech underscores the need for appropriate legal measures to regulate and address this issue. Given the complexities involved in balancing the fundamental rights and principles, especially the broad scope of freedom of expression established in Article 19(a) of the Indian Constitution, it becomes imperative

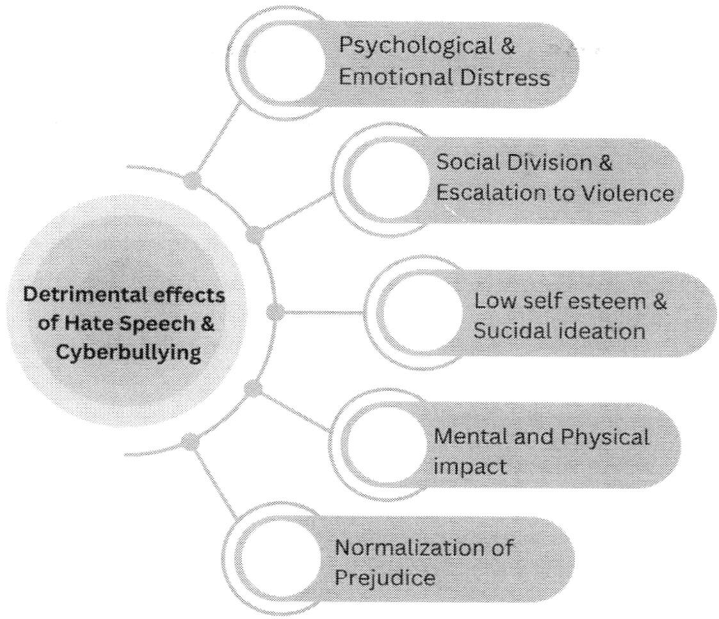

FIGURE 7.2 Detrimental effects of hate speech and cyberbullying.

to establish a legal framework that can effectively tackle hate speech. The goal of such laws should be to strike a careful balance between safeguarding the right to free speech and promoting the welfare and harmony of individuals and communities online. The provisions that establish penalties and remedies against hate speech are as follows:

According to Indian Penal Code Section 153A, Section 153A prohibits individuals from promoting disharmony or enmity between various social groups based on factors like religion, race, language, or region through spoken or written words, actions, or any other means. Those found guilty could be sentenced to imprisonment for a maximum of three years, fined, or both (Indian Penal Code, 1860). According to Section 295(A) of the Indian Penal Code, anyone who deliberately and maliciously seeks to offend the religious sentiments of any group of citizens of India, whether through spoken or written words, signs, visible representations, or any other means, by insulting or attempting to insult their religion or religious beliefs, shall be subject to imprisonment for up to three years, a fine, or both (Indian Penal Code, 295[A]).

The Indian Penal Code's Section 509(B) addresses engaging in sexual harassment through electronic means, such as telecommunications devices or the internet, by generating, soliciting, or initiating the transmission of any comment, request, suggestion, proposal, image, or other communication that is obscene, lewd, lascivious, filthy, or indecent with the intention to harass or with the knowledge that it would cause harassment, annoyance, or mental distress to a woman. This is punishable by imprisonment for a duration not less than six months, but which may extend to two years, along with a fine (Indian Penal Code, 509[B]).

7.3.2 Legal and Ethical Considerations in Cyberbullying Regulation

Despite the absence of specific legislation in India targeting cyberbullying directly, certain provisions within existing laws are relevant to addressing various forms of cyberbullying. The Information Technology Act is a noteworthy provision, although its effectiveness is limited in the face of the escalating incidence of online crimes. The primary objective of the Information Technology Act was primarily focused on regulating e-commerce activities, as evidenced by its preamble. Nevertheless, the judiciary has interpreted and applied this legislation to encompass a broader range of internet-related concerns, including cyberbullying, cyberstalking, and similar offenses.

The IT Act of 2000, amended in 2008, is a legislation enacted by the Indian Government to specifically address cybercrimes and establish penalties for those offenses. This comprehensive law defines different types of cybercrime and the accompanying punishments for each act. While cyberbullying is recognized as a harmful act with long-lasting consequences, it is unfortunate that there is no specific provision within the Act solely dedicated to tackling this issue. The absence of a concrete rule to combat cyberbullying in India is a concerning matter, especially considering the rising incidence of such incidents. It is worth noting that the 2013 amendment to the Act introduced the criminal offense of cyberstalking, but the inclusion of cyberbullying remains pending. Nonetheless, certain clauses within the law's Chapter XI may provide some options for handling instances of cyberbullying:

- Section 66(A) addresses the penalties for posting or sending offensive, disparaging, abusive, or harmful statements or content on social media platforms, in group chats, or through other channels.
- Section 66(D) addresses the penalty for making use of a computer device to personate. Blackmailing or misleading another individual online or on social media platforms can result in prison time of up to three years and penalties of at least one lakh rupees.
- Section 66(E): This section discusses the consequences associated with the violation of privacy. If an individual unlawfully exploits someone's photographs or divulges personal information, they can be charged with cyberbullying. As per the provision outlined in this clause, the offender may be subject to a maximum of three lakh rupees in fines or imprisonment for a maximum of three years.

(Information Technology [Amendment] Act, 2008, § 66[E])

The implementation and enforcement of anti-cyberbullying laws are of paramount importance in today's digital age. These laws serve as a crucial protective measure for vulnerable individuals, particularly children and teenagers, who are often the primary victims of online harassment. Creating a legal framework that addresses cyberbullying, and its consequences sends a strong message that such behavior will not be tolerated. It is equally important to foster an environment where victims feel empowered and supported to report cybercrimes without fear of retaliation. Through comprehensive legislation and proactive enforcement, we can strive toward a safer and more inclusive digital landscape, free from the harmful effects of cyberbullying.

7.4 DETECTION APPROACHES AND TECHNIQUES

The detection and mitigation of hate speech and cyberbullying have become pressing concerns in today's digital age. As online platforms continue to expand, so does the potential for the dissemination of harmful and abusive content. Researchers and practitioners have recognized the need for effective detection techniques to combat these destructive behaviors. Within this literature survey, we delve into the current corpus of research concerning the detection of hate speech and cyberbullying, shedding light on diverse methodologies and progressions within the discipline.

Table 7.1 provides a summary of various detection approaches applicable to hate speech and cyberbullying tasks, along with their descriptions. This information aids in determining the most suitable method for addressing such tasks.

In this section, we will examine each of these detection approaches to gain a comprehensive understanding of them.

7.4.1 TEXT-BASED APPROACHES

Text-based techniques have found extensive usage in the identification of cyberbullying and hate speech. These approaches leverage natural language processing (NLP) techniques to analyze the textual content and identify instances of offensive language, hate speech, or cyberbullying. Features such as n-grams, linguistic patterns, and sentiment analysis are utilized to capture the underlying characteristics of abusive content. ML algorithms, including support vector machines (SVM), XGBoost, and neural networks, are then trained on labeled datasets to classify text as either abusive or non-abusive. In this section, we will understand the different ways in which the text-based approach was used for the detection task and discuss the important observations highlighted in the existing literature.

The researchers utilized a hate speech lexicon derived from Hatebase.org, which comprised words and phrases recognized by internet users as constituting hate speech (Davidson, Warmsley, Macy, & Weber, 2017). They collected a dataset by scraping tweets from the Twitter API and searching for tweets that contained terms from the lexicon. These tweets were subsequently manually labeled as either Hate, Offensive

TABLE 7.1
Summary of Different Detection Approaches

Detection Approaches	Description
Text-Based approach	Leverages features like n-grams, linguistic patterns, and sentiment analysis to capture the underlying characteristics of content.
Contextual features and User characteristics	Contextual elements along with user-specific information are important.
Multimodal approach	Visual cues like images, videos, and audio help build robust detection models.
Machine Learning and NLP techniques	enable automated analysis of vast amounts of text data, identifying patterns, and classifying instances of hate speech and cyberbullying.

but not hate, or neither offensive nor hate speech. Different methods were used to bring in dimensionality reduction. While the logistic regression model demonstrated favorable precision, recall, and F1 scores, it displayed a tendency to classify tweets as less hateful or offensive compared to evaluations made by human coders.

Lu, Nijia, et al. (2020) employed a text-based approach using the Char-CNN model to address the challenges associated with short, noisy, and unstructured social media text containing spelling errors. By leveraging character-level analysis, the model effectively handled spelling errors and unclear texts. Incorporating shortcuts and a focal loss function enabled the detection of nuanced bullying signals and addressed class imbalance issues. The model demonstrated competitive results compared to state-of-the-art techniques in experimental evaluations on Chinese Weibo and English Tweet datasets.

Additionally, Paul and Saha (2022) proposed CyberBERT, a specialized variant of BERT, for cyberbullying identification. Leveraging the power of BERT—a multi-layer bidirectional Transformer encoder pretrained on a large corpus of text—the model successfully represented single and paired sentences. Through knowledge distillation, the researchers compressed the information from a larger model into a smaller one, reducing computational costs. By fine-tuning BERT with the addition of a fully connected layer and a softmax classifier, the CyberBERT model recorded state-of-the-art results in cyberbullying classification. Thus, the study highlights the effectiveness of combining text-based approaches with deep learning models like Char-CNN and BERT. This combination can even handle noisy and unstructured data and has achieved competitive results while maintaining lower computational costs.

7.4.2 Contextual Features and User Characteristics

In recent years, researchers have recognized the importance of considering contextual features and user characteristics in hate speech and cyberbullying detection. Factors such as the identity of the speaker, the target of the abuse, and the platform on which the content is shared can significantly impact the interpretation and severity of abusive language. Incorporating these contextual features, along with user-specific information such as age, gender, and social network connections, has shown promise in improving the effectiveness of detection models. The work emphasizes two key claims that are supported by its results. Firstly, unlike the other approaches that assume cyberbullying instances primarily consist of vulgar or profane words, they demonstrate that negative language is not always present. Secondly, the authors emphasize the significance of considering contextual factors, such as the attributes of the individuals involved and their positions within the network structure, in order to enhance the efficacy of cyberbullying detection methods. The results of their research thus conclude that socio-contextual features improve the efficiency of cyberbullying detection.

Mosca, Wich, and Groh (2021) argue in their study that user characteristics play a crucial role in the decision-making process of hate speech detection models. Detecting hate speech is a highly challenging task, particularly when hate expressions are implicit and concealed within figures of speech and sarcasm. To address this complexity, the methodology incorporates techniques from the field of explainable AI

(XAI), enabling a better understanding of the model's behavior. The inclusion of user and social context in the models has shown improved performance and reduced bias from text alone. The study incorporates data from three different sources: the text of the tweets, the vocabulary used by the users, and the structure of the follower network. However, identifying sexism solely using a text model is more challenging compared to racism, despite the sexism class being twice as large. The Davidson and Waseem dataset was used for experimentation. The research emphasizes that accuracy alone should not be the sole metric for model preference, and data scarcity remains a challenge, especially when incorporating context features, which are often difficult to obtain.

Pérez, Juan Manuel, et al. (2023) conducted a study to see how context affects the detection of hate speech. They gathered user responses to tweets from major news outlets in Argentina and created a dataset. The dataset included not only the offensive responses but also the entire news article and its accompanying comments. Using transformer-based techniques, the researchers ran classification experiments to evaluate the impact of context. The results clearly showed that including context improved performance. The models achieved a significant 4–5 point spike in Macro F1 scores when context was added. With just a little context, that is, a tweet, the best results in their experiments were reported, and the researchers suggested exploring other sources of information in future work. They proposed incorporating real-world knowledge about the targets of hate speech. Thus, the results from the recent experiments have proven thattext alone is not sufficient for the detection task, adding context and user characteristic features increases the model's efficiency.

7.4.3 Multimodal Approaches

With the increasing prevalence of multimedia content, researchers have begun exploring multimodal approaches for hate speech detection. These approaches integrate textual, visual, and acoustic features to capture the full spectrum of abusive behavior. By considering not only the text but also visual cues such as images, videos, and audio, multimodal models offer a more comprehensive understanding of abusive data and enable robust detection.

In their paper, Rana Aneri, and Sonali Jha explore the concept of modality, which relates to how something is sensed or expressed (Rana & Jha, 2022). They highlight that humans see the world through various modalities, such as sound, images, language, etc. In the context of hate speech, individuals rely on a combination of audio, visual, and text data to conclude whether a specific instance is racist or sexist. Thus, they propose that ML models must consider data in various forms for effective identification of hate speech. The authors specifically focus on language as a modality and capture the semantics and paraverbal cues. They built a DL model that integrates text and emotion to classify hate speech, yielding competitive results. Thus, substantiating the advantages of integrating speech features instead of solely depending on text.

Cheng, Lu, et al., in their work, address the limitations of existing cyberbullying detection models that primarily rely on text analysis (Cheng, Li, Silva, Hall, & Liu, 2019). Recognizing that cyberbullying incidents on social media often involve multiple modalities, the study proposes a comprehensive approach that incorporates textual, visual, and temporal information. By leveraging deep learning techniques, the XBully system extracts meaningful features from user profiles, image attributes, time of posting, and textual content. The experiment utilizes two real-world social media datasets, including Instagram and Vine, to validate the effectiveness of the multimodal approach. The research contributes to a more holistic understanding of cyberbullying by considering the various modalities involved and emphasizes the significance of comprehensive detection methods in addressing the prevalence of cyberbullying. Thus, the findings imply that the model's performance is further enhanced by having various modalities and incorporating knowledge from diverse points of view.

7.4.4 Machine Learning and Natural Language Processing Techniques

One of the widely used approaches in the identification of Online Hate Speech and Cyberbullying is to make use of various ML and NLP techniques. The general pipeline of any NLP-based technique will follow the following steps:

A. Data Preprocessing: Machine learning algorithms often require preprocessing steps to clean and prepare textual data. Techniques such as tokenization (dividing text into single words), omission of stop words, and stemming (word in its base form) help in standardizing the input data and improving the efficiency of subsequent analysis.
B. Feature Extraction and Feature Selection: NLP techniques are employed to pick useful features from text data. This uses techniques like bag-of-words representation, where the frequency of occurrence of words in a text is used as a feature, or more advanced techniques such as word embeddings (e.g., Word2Vec or GloVe), which capture semantic relationships between words. These features serve as input for ML models. Once the features are extracted, feature selection is used to identify the most relevant subset of features and eliminate redundant features.
C. Model Training: Training involves feeding the labeled data to the ML/DL models, which can be further fine-tuned for the task at hand by adjusting their hyperparameters.
D. Model Evaluation: The performance of a model is assessed using evaluation metrics like precision, recall, F1 score, or accuracy, which indicate how well it can classify hate speech and cyberbullying instances.
E. Online Monitoring and Filtering: Trained models can be deployed in real-time applications to monitor online content, such as social media platforms or comment sections, to detect and filter out instances of hate speech and cyberbullying. Automated systems can flag potentially offensive content for manual review or take immediate action based on predefined rules or thresholds.

In addition to traditional ML models, researchers have incorporated various neural networks and state-of-the-art Transformer-based models like BERT (Bidirectional Encoder Representations from Transformers), for the detection task. Moreover, integrating AI techniques with knowledge management systems enhances their effectiveness (Girard & Girard, 2015). Knowledge management allows organizations to systematically capture data insights from the AI tools used for hate speech and cyberbullying detection. By documenting both successes and failures, KMS helps refine AI-driven models over time. This continuous feedback loop ensures that AI systems evolve based on real-world experiences, making detection tools more robust, context-aware, and adaptable to emerging threats.

7.5 CHALLENGES AND LIMITATIONS

7.5.1 Contextual Challenges in Hate Speech and Cyberbullying Detection

Detecting and accurately classifying hate speech and cyberbullying can be particularly challenging due to the contextual complexities involved. Here are some key contextual challenges researchers and practitioners face:

A. Evolving Language and Slang: Online communication is constantly evolving, with new slang terms, abbreviations, and cultural references emerging regularly. This presents a challenge for hate speech and cyberbullying detection systems, as they need to stay updated to effectively recognize and understand these linguistic variations. For instance, a study by Founta et al. (2018) highlighted the challenges posed by evolving language in online hate speech detection, emphasizing the need for continuous adaptation and monitoring of new linguistic trends.

B. Irony, Sarcasm, and Satire: Recognizing and interpreting nuanced forms of language, such as irony, sarcasm, and satire, is crucial in determining the intent behind a statement. For example, research by Albadi, Kurdi, and Mishra (2018) explored the detection of irony and hate speech in social media, emphasizing the need for context-aware approaches to accurately differentiate between harmful and clean speech.

C. Contextual Understanding: As mentioned in the previous section, understanding the context in which a particular statement is made is vital for accurate hate speech and cyberbullying detection. An isolated text snippet may not provide enough information to determine its intent. Recent research by Badjatiya, Gupta, Gupta, and Varma have developed a context-aware approach for hate speech detection, leveraging contextual features such as user information, conversation history, and social network connections to improve accuracy.

D. Cultural and Regional Variations: The interpretation and perception of hate speech can vary across different cultural and regional contexts. What is deemed offensive in one culture may be deemed acceptable in another. Addressing cultural and regional variations is a significant challenge in developing universal hate speech detection models. Research by Fortuna and Nunes (2018) explored the challenges of automatic hate speech detection across languages, highlighting the need for language-specific and culturally sensitive approaches.

E. Linguistic Variations and Code-switching: Multilingual online platforms and users who code-switch between languages pose challenges for hate speech detection systems. Understanding and analyzing hate speech in multilingual contexts requires language-specific models and techniques. For instance, research by Zampieri et al. focused on hate speech detection in code-switched social media text, highlighting the need for language-aware approaches to address these variations.
F. Memes and Visual Content: Hate speech and cyberbullying often manifest in the form of memes, images, or videos, which can be challenging for text-based detection systems. Analyzing visual content requires the integration of image recognition, sentiment analysis, and text-image correlation techniques.

These contextual challenges emphasize the complexities involved in accurately detecting and understanding hate speech and cyberbullying. Researchers and practitioners are continuously exploring novel approaches, including context-aware models, linguistic analysis, and cross-cultural considerations, to overcome these challenges and develop more effective detection systems.

7.5.2 Ethical Considerations and Potential Biases in Hate Speech and Cyberbullying Detection

While hate speech and cyberbullying detection systems aim to combat harmful online behavior, they can inadvertently introduce ethical concerns and biases. Automated systems rely on ML algorithms trained on large datasets, which can reflect existing biases present in society. If the training data is biased or incomplete, the detection algorithms may exhibit similar biases, potentially leading to discriminatory outcomes. Recent research by Davidson, Bhattacharya, and Weber (2019) demonstrated that hate speech detection models trained on publicly available datasets exhibited biases against certain political and social groups. The study highlighted the potential for biased training data to perpetuate discriminatory outcomes, leading to false positives or negatives in hate speech detection algorithms. Striking a balance between effectively detecting harmful content and minimizing biases is a complex task that requires continuous monitoring and improvement.

Moreover, there are ethical considerations regarding the privacy and consent of individuals involved in hate speech and cyberbullying incidents. A study by Sap et al. investigated the unintended consequences of using ML models for detection tasks. It revealed that these models often exhibited racial, gender, and regional biases, leading to disproportionately flagging content from marginalized communities. This raises concerns regarding fairness and the potential for discriminatory treatment. Care must be taken to ensure that the detection methods and actions taken to address these issues do not infringe upon individuals' privacy rights or subject innocent individuals to unwarranted scrutiny.

Additionally, distinguishing between hate speech and legitimate freedom of expression can be challenging. Recent studies have focused on addressing the delicate balance between protecting individuals from harm and preserving free expression rights. For example, Bender, Gebru, McMillan-Major, and Shmitchell (2019) explored the ethical considerations of hate speech detection, emphasizing the need for transparency and accountability in algorithmic decision-making to avoid unnecessary censorship and uphold freedom of speech. The fine line between offensive but legally protected speech and hate speech requires careful consideration, as misclassifying content can impede free expression rights. Addressing these ethical considerations and potential biases requires a multidisciplinary approach, involving input from linguists, ethicists, psychologists, and experts in ML and artificial intelligence. Regular audits, transparency, and user feedback mechanisms are crucial to mitigate potential biases and ensure the fairness and effectiveness of hate speech and cyberbullying detection systems.

The detection of hate speech and cyberbullying poses significant contextual challenges that necessitate attention from researchers and practitioners. These challenges encompass evolving language and slang, the intricate nature of irony and sarcasm, the crucial role of contextual understanding, cultural and regional variations, linguistic variations and code-switching, as well as the analysis of visual content such as memes. To overcome these challenges, the development of context-aware models, linguistic analysis techniques, and cross-cultural considerations is imperative. Additionally, ethical considerations and potential biases emerge in the realm of identifying cyberbullying and hate speech. Transparency, accountability, regular audits, and user feedback mechanisms play vital roles in mitigating biases and ensuring fairness and effectiveness in detection systems. By actively tackling these contextual challenges, ethical considerations, and potential biases, researchers and practitioners can forge ahead in the creation of more precise, inclusive, and equitable hate speech and cyberbullying detection systems. Such endeavors are critical in fostering online safety, safeguarding individuals' mental well-being, and upholding fundamental rights within the digital domain.

7.6 CASE STUDIES

Reddit has integrated machine learning (ML) and artificial intelligence (AI) to combat online harassment and hate speech across its platform. In 2021, as mentioned in the Reddit post of enthusiastic potato (2024), Reddit introduced a new harassment filter and user reporting mechanism aimed at enhancing content moderation. The ML-driven harassment filter detects abusive behavior by analyzing user comments and post patterns, flagging harmful content for review by human moderators. Reddit's AI tools assist in identifying hate speech by learning from user reports and moderator actions, allowing the system to adapt over time. The use of these tools enables Reddit to maintain community guidelines while balancing free speech with the need to protect users from harmful content. Meta has developed advanced AI systems to automatically detect and moderate hate speech on its platforms, including Facebook and Instagram. A key tool in their arsenal is the Few-Shot Learner (FSL), an AI model capable of identifying new and evolving types of hate speech with minimal training data. FSL has proven effective in detecting hate speech in multiple languages and adapting quickly to changes in harmful content trends. As per Schroepfer (2021), Meta reported that their AI systems proactively detected and removed 97%

of hate speech before users even flagged it. This demonstrates the scalability and efficiency of ML in content moderation, reducing human oversight for harmful content while improving safety across Meta's platforms.

Flannery O'Connor and Moxley (2023) said YouTube employs AI and ML technologies to detect and remove hate speech and other harmful content in videos and comments. Their system is based on NLP and pattern recognition models that can identify inappropriate content across different languages and contexts. YouTube's AI-powered moderation tools flag content for review by human moderators and automatically take down flagged videos that violate community guidelines. In their blog, YouTube emphasizes responsible AI innovation, striving for transparency, fairness, and effectiveness in moderation, with models improving continuously through feedback loops. The platform has also been refining its ability to reduce the reach of borderline content without infringing on free speech. These case studies show the application of ML to detect and moderate hate speech on widely used social media platforms.

7.7 FUTURE DIRECTIONS AND CONCLUSION

As we continue to address the pressing issue of hate speech and cyberbullying, there are several areas that warrant further exploration and development. Firstly, advancements in ML techniques, particularly the integration of multimodal analysis combining text, images, and user behavior, hold promise for more robust detection of hate speech and cyberbullying. This holistic approach can provide a deeper understanding of the context and intent behind harmful content. However, we must also acknowledge the challenges that persist in this domain. The ever-evolving landscape of slang, memes, and user-generated content necessitates the continuous adaptation of detection strategies to stay abreast of emerging trends.

Mitigation strategies play a crucial role in combating hate speech and cyberbullying. Automated content moderation tools, powered by ML algorithms, can assist in the identification and removal of harmful content, thereby fostering safer online environments. User reporting mechanisms also empower individuals to report instances of hate speech and cyberbullying, allowing for swift action to be taken. Furthermore, education and awareness campaigns are instrumental in promoting digital literacy, empathy, and responsible online behavior, fostering a culture of tolerance and respect. In conjunction with technological advancements and mitigation strategies, legal frameworks must also be strengthened to address hate speech and cyberbullying. Legislation should clearly define and criminalize these acts while ensuring appropriate penalties for offenders. Collaborative efforts between government bodies, law enforcement agencies, technology companies, and civil society are essential to effectively enforce these laws and create a safer online ecosystem.

In conclusion, combating hate speech and cyberbullying requires a multi-faceted approach that encompasses technological innovation, mitigation strategies, education, and legal measures. By continuing to refine detection methods, addressing the challenges faced in identifying hate speech, and promoting responsible online behavior, we can work toward a more inclusive and respectful digital space for all.

REFERENCES

Albadi, N., Kurdi, M., & Mishra, S. (2018). Are They Our Brothers? Analysis and Detection of Religious Hate Speech in the Arabic Twittersphere. In *Proceedings of the 2018 IEEE/ACM International Conference on Advances in Social Networks Analysis and Mining (ASONAM)*, 69–76. https://doi.org/10.1109/ASONAM.2018.8508247

BBC News. (15 Nov. 2021). Online Hate Speech Rose 20% During Pandemic: 'We've Normalised It'. *BBC News*. www.bbc.com/news/newsbeat-59292509

Bender, E. M., Gebru, T., McMillan-Major, A., & Shmitchell, S. (2019). On the Dangers of Stochastic Parrots: Can Language Models Be Too Big? In *Conference on Fairness, Accountability, and Transparency*, 610–623. https://doi.org/10.1145/3442188.3445922

Broadband Search. (2023, March 15). All the Latest Cyberbullying Statistics for 2023. www.broadbandsearch.net/blog/cyber-bullying-statistics#post-navigation-1

Cheng, L., Li, J., Silva, Y., Hall, D., & Liu, H. (2019). XBully: Cyberbullying Detection Within a Multi-Modal Context. In *Proceedings of the 12th ACM International Conference on Web Search and Data Mining (WSDM)*, 339–347. https://doi.org/10.1145/3289600.3291037

Davidson, T., Bhattacharya, D., & Weber, I. (2019). Racial Bias in Hate Speech and Abusive Language Detection Datasets. In *Proceedings of the 2019 Conference of the North American Chapter of the Association for Computational Linguistics: Human Language Technologies*, Volume 1 (Long and Short Papers), 25–35. https://doi.org/10.18653/v1/W19-3504

Davidson, T., Warmsley, D., Macy, M., & Weber, I. (2017). Automated Hate Speech Detection and the Problem of Offensive Language. In *Proceedings of the International AAAI Conference on Web and Social Media*, 512–515. https://doi.org/10.1609/icwsm.v11i1.14955

Diaz, J. (2021, May 10). Social Media Hate Speech, Harassment 'Significant Problem' For LGBTQ Users. *NPR*. https://www.npr.org/2021/05/10/995328226/social-media-hate-speech-harassment-significant-problem-for-lgbtq-users-report

enthusiastic-potato[@u/enthusiastic-potato]. (2024, March 12). Reddit: A New Harassment Filter and User..reporting..type. https://www.reddit.com/r/modnews/comments/1bd3b82/a_new_harassment_filter_and_user_reporting_type/?share_id=wejmZAGqJN4vs_l--OrMC&rdt=49804

Flannery O'Connor, J. & Moxley, E. (2023, November 14). YouTube: Our Approach to Responsible AI Innovation. https://blog.youtube/inside-youtube/our-approach-to-responsible-ai-innovation/

Fortuna, P., & Nunes, S. (2018). A Survey on Automatic Detection of Hate Speech in Text. *In ACM Computing Surveys*, 51(4), 1–30. https://doi.org/10.1145/3232676

Founta, A., Djouvas, C., Chatzakou, D., Leontiadis, I., Blackburn, J., Stringhini, G., Vakali, A., Sirivianos, M., & Kourtellis, N. (2018). Large Scale Crowdsourcing and Characterization of Twitter Abusive Behavior. In *Proceedings of the International AAAI Conference on Web and Social Media*, Vol. 12, No. 1. https://doi.org/10.1609/icwsm.v12i1.14991

Girard, J., & Girard, J. (2015). Defining Knowledge Management: Toward an Applied Compendium. *Online Journal of Applied Knowledge Management*, 3, 1–20.

Hamm, M. P., Newton, A. S., Chisholm, A., Shulhan, J., Milne, A., Sundar, P., Ennis, H., Scott, S. D., & Hartling, L. (2015). Prevalence and Effect of Cyberbullying on Children and Young People: A Scoping Review of Social Media Studies. *JAMA Pediatrics*, 169(8), 770–777. https://doi.org/10.1001/jamapediatrics.2015.0944

Hinduja, S., & Patchin, J. W. (2017). Cultivating Youth Resilience to Prevent Bullying and Cyberbullying Victimization. *Child Abuse & Neglect*, 73, 51–62. https://doi.org/10.1016/j.chiabu.2017.09.010

Information Technology (Amendment) Act. (2008). § 66(E), No. 10, Acts of Parliament, 2009.

Lu, N., Wu, G., Zhang, Z., Zheng, Y., Ren, Y., & Choo, K.-K. R. (2020). Cyberbullying Detection in Social Media Text Based on Character-Level Convolutional Neural Network With Shortcuts. *Concurrency and Computation: Practice and Experience*, 32. https://api.semanticscholar.org/CorpusID:213465177

Mosca, E., Wich, M., & Groh, G. (2021). Understanding and Interpreting the Impact of User Context in Hate Speech Detection. In *Proceedings of the Ninth International Workshop on Natural Language Processing for Social Media*, 91–102. https://doi.org/10.18653/v1/2021.socialnlp-1.8

Patchin, Justin. (2019). 2019 cyberbullying data. Accessed July 30, 2019, https://cyberbullying.org/2019-cyberbullying-data

Paul, S., & Saha, S. (2022). CyberBERT: BERT for Cyberbullying Identification. *Multimedia Systems*, 28(6), 1897–1904. https://doi.org/10.1007/s00530-020-00710-4

Pérez, J. M., Gravano, A., Luque, F. M., Zeyat, D., Kondratzky, M., Moro, A., Serrati, P. S., Zajac, J., Miguel, P., Debandi, N., & Cotik, V. (2023). Assessing the Impact of Contextual Information in Hate Speech Detection. *IEEE Access*, 11, 30575–30590.

Rana, A., & Jha, S. (2022). Emotion Based Hate Speech Detection Using Multimodal Learning. arXiv preprint arXiv:2202.06218.

Schroepfer, M. (2021, February 21). Update on Our Progress on AI and Hate Speech Detection. *Meta*. https://about.fb.com/news/2021/02/update-on-our-progress-on-ai-and-hate-speech-detection/

8 Combatting Cyberthreats in African Digital Space with Artificial Intelligence

Basil Osayin Daudu and Goddy Uwa Osimen

8.1 INTRODUCTION

Anyone can fall prey to the nefarious activities of cybercriminals if not properly guided. A lot of people and companies have been defrauded as some affected persons have committed suicide and some affected companies have folded up because of being defrauded by cybercriminals. What affects one affects the other and what affects one aspect of society affects the others. As such, no sector of society is completely free from the activities of cybercriminals. Cybercrime in itself is deadly, and its impact is fierce. African countries are enthusiastically adopting digital advancements, incorporating technology across multiple sectors such as education, sports, military operations, healthcare, agriculture, and governance. Currently, there is a swift expansion of the digital landscape in Africa, promising notable impacts on economic growth, societal empowerment, and technological progress. However, this shift toward digitalization also brings a rise in cybercrime, posing significant hurdles to Africa's digital ambitions.

Cybercrime continues to escalate despite existing cybersecurity measures, with African leaders engaged in prolonged battles against it. Regrettably, cyber threats contribute to security breaches, posing risks to national security through various criminal activities facilitated by the internet. Cybercrime encompasses a multitude of illicit actions carried out through online channels. Exploiting vulnerabilities in computer systems and networks, cybercriminals gain unauthorized access, pilfer sensitive information, disrupt services, and inflict financial or reputational harm on individuals, organizations, and governments (Lehman & Shirelle, 2005). According to Zola (2023), in an era of rapid digital inclusion and increasing technology integration into daily life, the ramifications of cybercrime are extensive and varied. As noted by Welance (2023), the consequences of cybercrime can be catastrophic, spanning from financial loss and identity theft for individuals to reputational damage and legal repercussions for businesses. Ajijola and Allen (2022) reported that progress in developing and implementing national-level cybersecurity strategies in Africa has been limited.

DOI: 10.1201/9781003498094-8

The key aspects of these strategies include their design, implementation, and impact. Effective cyber strategies must, at a minimum, explain their necessity, and outline actions, timelines, responsibilities, and funding and implementation methods. Only three African countries—Eswatini, Kenya, and Senegal—meet these essential criteria, which include a threat assessment, a plan of action, a timeline, assigned responsibilities among key stakeholders, and clear resource allocation provisions.

Furthermore, cybercrime can exert profound effects on society, spanning economic repercussions, national security considerations, and a surge in cyberbullying and harassment. Tahiru (2018) underscores the necessity for a unified strategy and cybersecurity frameworks among African Union Member States. This notable rise emphasizes the urgency for heightened awareness and preventative measures to mitigate cyber threats. Therefore, this chapter contends that African nations must fortify their digital sphere and safeguard it against cybercrime to nurture sustainable development and global solidarity. To achieve this objective, pragmatic steps such as enacting legislation and leveraging artificial intelligence (AI) applications and innovations should be prioritized.

8.2 THEMATIC REVIEW OF CYBERCRIME

African digital space, in as much as its benefits to the African nations are remarkable, and one of its nightmares is cybercrime. According to Kundi and Akhtar (2014), cybercrime stands as one of the most significant global challenges, impacting nations both developed and developing, and countries are actively combating cybercriminals through existing legislation. The effective monitoring, control, and prevention of cybercrimes necessitate cyber-legislation. Nations worldwide are grappling with the perils of cybercrimes due to various factors, including inadequate technology, institutional incapacity, absence of legal frameworks, financial constraints, and insufficient cooperation with international law enforcement agencies. Cybercrime encompasses a spectrum of illicit activities conducted via the internet, such as breaching internet security, spoofing or unauthorized access to email accounts, unauthorized access to bank account passwords, and spyware attacks, all of which carry severe penalties (Kundi & Akhtar, 2014).

Halder and Jaishankar (2011) characterize cybercrimes as offenses perpetrated by individuals or groups with criminal intentions to cause intentional harm, whether physical or mental, directly or indirectly, to individuals, groups, or organizations. These crimes utilize telecommunication networks such as chat rooms, emails, notice boards, groups, and mobile phones for SMS/MMS communication. Cybercrime, or computer crime, can be categorized into three main types: first, using a computer as a target to attack other computers by infecting them with viruses or spreading malware; second, using a computer as a weapon to commit traditional crimes such as fraud or illegal gambling; and third, using a computer as an accessory by storing illegal or stolen information or data (Mativat & Tremblay, 1997). Additionally, Chaubey (2012) regards the internet as a double-edged sword with both advantages and disadvantages, cybercrime being one of the disadvantages. Chaubey (2012) asserts that cybercrime is an uncontrollable menace rooted in the misuse of the increasing reliance on computers in modern life. The widespread use of computers and related technology in daily life is rapidly growing and has become an essential element that facilitates user convenience. However, despite the benefits the internet offers, it also harbors dark sides.

On his part, Viswanathan (2001) gave three definitions of cybercrime: Any illegal action in which a computer is the tool or object of the crime, that is, any crime, the means or purpose of which is to influence the function of a computer. (ii) Any incident associated with computer technology in which a victim suffered or could have suffered loss and a perpetrator, by intention, made or could have made a gain. (iii) Computer abuse is considered any illegal, unethical, or unauthorized behavior relating to the automatic processing and transmission of data. Lehman and Shirelle (2005) highlight that cybercrime presents notable obstacles for law enforcement officials due to its nonlocal nature and clandestine execution, frequently transcending geographical boundaries. Cybercrime, also known as computer crime, e-crime, hi-tech crime, or electronic crime, typically denotes criminal activities wherein a computer or network serves as the source, tool, or target of the offense. Tubake (2013) contends that the notion of cybercrime does not fundamentally differ from that of conventional crime, as both entail breaches of law and order. Thus, Tubake (2013) outlined several distinctive characteristics of cybercrimes:

i. The primary tool used in cybercrimes is technology. These offenses are executed through technology, and consequently, cybercriminals are individuals well-versed in the workings of the internet and computers.
ii. Cybercrime is highly efficient, often occurring in real time. It can take mere seconds or minutes to hack websites or perpetrate cyber fraud.
iii. Cybercrime transcends geographical boundaries and limitations. A cybercriminal situated in one corner of the world can carry out hacking activities on systems located in distant corners of the globe.
iv. The commission of cybercrimes occurs within cyberspace, allowing cybercriminals to operate from physical locations outside of cyberspace. All aspects of cybercriminal activity, from preparation to execution, occur within the realm of cyberspace.
v. Cybercrime has the potential to cause extensive harm and damage of unimaginable proportions. It can swiftly dismantle websites created and maintained with substantial investments or breach the security of banks and defense department websites, resulting in significant losses.
vi. Collecting evidence of cybercrime and proving it in a court of law is exceptionally challenging due to the anonymity and invisibility of cybercriminals. Additionally, cybercrimes have the potential to impact multiple countries simultaneously, further complicating legal proceedings.

As earlier stated, cybercrime is an off-shoot of the use of the internet, digital technology, and cyberspace. The term "cyberspace" originated from the science-fiction novel Neuromancer by William Gibson, which portrayed his conception of a three-dimensional realm consisting of pure information, traversing between various computer and computer clusters that form this expansive landscape (Kizza, 2014). The infrastructure of cyberspace encompasses hardware nodes responsible for sourcing, transmitting, and receiving data; software components such as protocols; human users interacting with information; and the information itself, which can exist in either a static state at a node or a dynamic state within linking media (Kizza, 2014).

Simply put, cyberspace is a network between computer systems. The digital environment in Africa serves as a miniature representation of the worldwide cyberspace, facilitating interactions, activities, experiences, and associations among both young and adult populations. However, the lack of regulation in these interactions raises concerns (Daudu et al., 2023). The African digital sphere is enhanced by the establishment of the African Digital Rights Network, which was founded in 2020 and is based in Brighton, East Sussex, United Kingdom. This network aims to provide an online platform for democratic expression, safeguarding privacy and freedom of speech. Its key objectives include expanding digital access, promoting social media activism, developing civic technologies, advocating for legislation that enhances digital rights, and advancing digital security technologies (Roberts, 2021). The African digital space, therefore, accommodates a multitude of users, including individuals, organizations, and government entities, alongside various digital opportunities and closures, which encompass political, social, and economic report documents. In this chapter, therefore, cybercrime is taken to mean any unlawful online means of livelihood, and a cybercriminal is seen as an irresponsible miscreant.

8.3 THE WORLD OF CYBERCRIMINALS: WHO THEY ARE AND WHAT THEY CAN DO

Determining the precise count of cybercriminals worldwide presents significant challenges, as many individuals have embraced cybercrime as a profession due to its lucrative nature. According to Toulas (2022), the hacker group OPERA1ER used readily available hacking tools to steal at least \$11 million from African banks and telecommunications providers. Between 2018 and 2022, they executed over 35 successful attacks, with approximately a third occurring in 2020. Other notable hacker groups include Witchetty, Mustang Panda, Daggerfly, Alloy Taurus, and MuddyWater. However, scholars have identified several distinct categories of cybercriminals, including hackers, identity thieves, insiders, organized crime groups, nation-states, and cyber terrorists (Cyber Talent, 2023; Mishra, 2023; Borges, 2024).

Hackers are skilled individuals who exploit their expertise to gain unauthorized access to computer systems or networks, often with the intent of stealing data or causing harm. Their motivations can range from financial gain to political or social causes, or simply a desire to test their abilities. Insiders, on the other hand, are individuals with legitimate access to computer systems or networks who misuse their privileges for personal gain or to undermine the organization. Identity thieves constitute another significant group of cybercriminals, specializing in stealing personal information for financial gain. Their tactics may involve using stolen identities to conceal their own identity or perpetrate various fraudulent activities, including credit card fraud and impersonation. The impact of identity theft on victims can be profound, leading to substantial financial losses and enduring emotional distress. The ease with which identity theft can be carried out, often facilitated by readily available personal data online, underscores the importance of robust data protection measures and individual vigilance.

Insiders encompass employees, contractors, or other trusted individuals who exploit their authorized access to steal data, disrupt systems, or partake in other illicit activities. Organized crime groups consist of individuals engaging in cybercrime primarily for financial gain or other criminal endeavors. These groups often operate with high levels of sophistication, employing advanced methods to evade detection and obscure their activities. Nation-states denote governments or state-sponsored entities utilizing cyber-attacks for political or military objectives, and such actions may involve intelligence gathering, system disruption, or seeking strategic advantages over rival nations. Cyber terrorists represent the most perilous faction of cybercriminals, utilizing cyberspace to perpetrate acts of terror, frequently targeting critical infrastructure and networks. Their motivations can span from political to ideological, to instill fear and chaos. Recent years have witnessed several prominent cyber terrorism incidents, including attacks on Iranian gas facilities and major cyber assaults against Danish power grids, underscoring the potential scale and severe ramifications of such attacks. Cyberterrorists may draw motivation from various political or social causes and employ diverse techniques to execute their agendas.

Furthermore, as noted by Cyber Talent (2023), some methods employed by cybercriminals include: (i) Spreading Malware through various means such as Google Forms, and software downloads, and (ii) Social engineering, which involves using psychological tactics to deceive individuals into divulging confidential information. (iii) Exploiting vulnerabilities for Remote Access, which may involve exploiting weaknesses in software or hardware, targeting IoT devices, or manipulating users into granting remote access under the guise of technical support. (iv) SQL Injection, a technique utilized to exploit weaknesses in databases to pilfer or alter information. (v) Cross-site scripting (XSS), a type of security vulnerability enabling attackers to insert malicious code into websites. (vi) Man-in-the-middle attacks, where the attacker intercepts and potentially alters communication between two parties. And the primary motivations behind cybercrime, so to speak, include financial gain, espionage, political or ideological motives, personal agendas, opportunism, and lack of awareness, among others (Mishra, 2023).

8.4 CYBERCRIME: TYPES, IMPLICATIONS, AND SOLUTIONS VIA ARTIFICIAL INTELLIGENCE

The predominant and most urgent cyber threat reported by African member countries is online scamming, which aims to exploit victims' fears, insecurities, and vulnerabilities through techniques such as phishing, mass mailing, and social engineering. Member countries have observed a significant rise in online banking scams, including incidents of banking and credit card fraud. (INTERPOL Report, 2021). Cybercrime has spread its tentacles to every nook and cranny of every African nation depriving nations and people of a better life. Inadequate information security infrastructure is seen as a major issue for Africa. In this regard, The Economic Commission for Africa (2022) reported that approximately 90% of African businesses are susceptible to cyberthreats. Currently, 28 out of 54 African countries have enacted data protection laws, while six more are in the process of developing such legislation. This lack of cybersecurity readiness is costing African Member States, on average, 10% of their GDP. According to Positive Technologies (2023), from early 2022 to mid-2023, the financial sector was the most frequently targeted by cyber-attacks, followed by telecommunications agencies, trade organizations, and the industrial sector. Cybercrime manifests in various forms, including but not limited to: phishing, hacking, malware dissemination, identity theft, cyberbullying, business email compromise, distributed denial-of-service (DDoS) attacks, and cyber extortion.

8.4.1 Phishing

This is a prevalent cybercrime that poses risks of identity theft and financial losses. Preventive measures against phishing attacks include using security software, activating multifactor authentication, regularly backing up data, and exercising caution with suspicious emails or messages (Ganacharya, 2020). In the event of suspecting a phishing attempt, it is crucial to verify the legitimacy of the communication before taking any action to safeguard personal information and accounts. Ganacharya (2020) suggests that identity theft can occur without hacking, as criminals also utilize social engineering tactics such as phishing to deceive individuals into voluntarily providing information.

Phishing encompasses two main forms: Email-based and pharming. In the email-based method, perpetrators undertake three phases. Firstly, they identify legitimate companies offering online services and communicating with customers electronically. Secondly, they create websites resembling legitimate ones, known as "spoofing sites," and often send emails resembling those from genuine companies to direct users to these sites. These sites then prompt users to disclose sensitive information like passwords. In the third phase, perpetrators use the obtained information to access the victim's accounts and carry out illegal activities, such as money transfers, often involving financial agents to launder the acquired funds (Abasilim et al., 2019). The second technique, pharming, involves manipulating the Domain Name Server (DNS) to direct users to spoofed websites. This typically involves installing malware on individual users' computers or DNS servers, redirecting users attempting to access their online accounts to fake sites. When users enter their personal information on these sites, identity thieves capture and record it.

8.4.2 Hacking

The concept "hacking" refers to the unauthorized access of a computer system, representing one of the earliest computer-related crimes that has evolved into a widespread phenomenon in recent years. By targeting computer systems containing extensive databases, perpetrators can acquire large-scale identity-related data, a method that has gained popularity. Hacking involves unauthorized entry into a computer system and stands as the most prevalent form of cybercrime globally. It is commonly defined as the act of breaking into computer systems. As a cybercrime, hacking poses significant risks to the internet, eroding its credibility by creating a perception of vulnerability and weakness.

According to Tubake (2013), there are four primary types of hacking prevalent today: (i) Hobbyist hacking, often undertaken by teenagers engrossed with the internet for entertainment purposes. (ii) Competitive hacking aimed at disrupting the business operations of rivals. (iii) Hacking with the intent of committing additional offenses such as fraud and misappropriation. (iv) Ethical hacking conducted by internet security firms to evaluate the security measures of their clients.

8.4.3 Malware

Instead of relying on email scams, identity thieves have turned to malicious software to directly obtain desired information from users' computers. As reported by INTERPOL (2021), the threat of ransomware is on the rise across the African continent, with reports indicating that over 61% of companies in the region fell victim to ransomware attacks in 2020 alone. These attacks targeted critical infrastructure in some African countries, including the healthcare and maritime sectors. Toolkits facilitating the creation of such malware are available for purchase at prices ranging from US$200 to US$500. Once installed, these software tools can operate in various ways. They can scan hard drives directly to extract information typically formatted as passwords, credit card numbers, and social security numbers. Given that many people use the same passwords for multiple accounts, uncovering just one or two passwords can provide access to all the user's accounts.

Additionally, malware can intercept communications or log keystrokes, essentially recording every input made by a user. This captured information can then be electronically sifted through for passwords and related data. The installation of malware involves the introduction of malicious software into a system, potentially causing harm or stealing sensitive data. Malware can be installed through various methods, including phishing emails, malicious websites, or exploiting software vulnerabilities (Zola, 2023). Malware, therefore, stands as one of the most prevalent tools used by cybercriminals, involving the covert installation of malicious software on victims' devices, ideally without detection. There exist various types of malwares, ranging from remote access Trojans that grant hackers control over devices to ransomware that holds devices hostage until a ransom is paid.

8.4.4 Identity Theft

Identity theft, also referred to as "yahoo yahoo," is a significant offense where an individual illicitly uses another person's personal or financial details without consent to perpetrate fraud, commonly observed among African youth. This criminal activity can result in financial losses, damage to one's credit history, and substantial expenditure of time and resources. Several warning signs indicate potential identity theft, such as receiving invoices for unauthorized purchases, debt collection calls for accounts not opened by the individual, or rejection of loan applications. According to Hussain (2004), safeguarding oneself involves refraining from sharing personal information like bank accounts or social security numbers with unfamiliar individuals. Additionally, regularly monitoring credit cards and bank statements for unauthorized transactions and understanding the methods used by thieves to obtain personal information are crucial in preventing identity theft.

Hussain (2004) further explains that when an individual's personal details are obtained from a compromised bank account, social media platform, or other online sources, hackers may open new financial accounts under the victim's name and exploit them to their limit before the victim becomes aware of any suspicious activity on their credit report. This type of fraud is prevalent enough to warrant the existence of identity theft protection services dedicated to combatting it. It's important to note that not all instances of identity theft are cybercrimes; some identity thieves acquire a person's information by physically stealing their mail, for instance.

8.4.5 Cyberbullying

Cyberbullying entails the use of digital technologies, such as social media or messaging platforms, to harass others. This behavior can encompass offensive name-calling, spreading false rumors, sharing explicit images without consent, making physical threats, and persistently probing about one's whereabouts. Research cited by Hussain (2004) indicates that nearly half of African teenagers have experienced cyberbullying, with older teenage girls being more susceptible targets. Cyberbullying poses a significant threat with detrimental effects on individuals, underscoring the importance of recognizing and addressing this form of harassment. Many African countries are working to introduce legal measures and sanctions against cyberbullying in their legislative frameworks. Cyberbullying can involve activities such as harassment and stalking, which are deemed illegal in various jurisdictions. Moreover, perpetrators may exploit cyberbullying to extract information or money from their victims.

Cyberbullying is marked by anonymity, with perpetrators often hiding their identity while engaging in activities aimed at uncovering sexually inappropriate or morally questionable information (Shariff, 2009). Similarities in aggressive behavior, power dynamics, and repetitive actions can be observed in both cyberbullying and traditional bullying (Hunter et al., 2007). Campbell (2015), as cited by Shariff (2009), describes cyberbullying as forms of bullying utilizing technology, a phenomenon increasingly adopted by children and adolescents to inflict harm on others. Cyberbullying can be perpetrated without physical contact with the victim, merely requiring access to an electronic device to execute intentions of causing harm (King, 2006). On his part, Aftab (2006) distinguishes between direct cyberbullying and cyberbullying by proxy (indirect). Direct cyberbullying involves sending messages directly to the victim, while cyberbullying by proxy entails the cyberbully enlisting others to carry out harmful actions, sometimes without their knowledge. Cyberbullies may also change the victim's account password to prevent access. Moreover, the key distinction between playful teasing and cyberbullying lies in the aggressive intent characteristic of cyberbullying.

8.4.6 Business Email Compromise

Business Email Compromise (BEC) stands as an intricate scam targeting both individuals and businesses, with the aim of illicitly acquiring money or vital information by deceiving them through email. Scammers often masquerade as trusted entities, soliciting payments for fictitious invoices or sensitive data. BEC scams manifest in various forms, such as data breaches, fraudulent invoice schemes, CEO impersonation, lawyer impersonation, and compromised accounts. According to the

FBI (2021), implementing preventative measures is crucial to thwart BEC attacks. Organizations can adopt strategies like utilizing secure email solutions, implementing multifactor authentication, providing employees with training to recognize warning signs, configuring security defaults, and utilizing email authentication tools. It is imperative to exercise caution, verify unusual requests, refrain from deviating from standard payment channels, and be vigilant for language anomalies or unusual date formats in emails to guard against BEC scams.

Furthermore, educating employees to trust their instincts and empowering them to question suspicious requests can help fortify defenses against BEC attacks. It is undeniable that nearly all online accounts are linked to an email address, providing cybercriminals with significant leverage if they gain access to your inbox (Gberevbie et al., 2014). With unfettered access, a thief could clandestinely peruse your emails, harvesting personal information. Alternatively, a hacker might alter the password and lock you out of your inbox, thereby gaining access to accounts linked to that email.

8.4.7 DDoS Attacks

DDoS attacks, which stand for Distributed Denial-of-Service attacks, are malicious efforts to disrupt regular traffic to a website by inundating a server with internet traffic, rendering it inaccessible to legitimate users. These attacks can have various motives, ranging from individuals seeking to exploit cyber vulnerabilities to financially driven attacks aimed at disrupting online operations for competitive advantage or extortion purposes. DDoS attacks have been increasing in frequency, impacting even major global companies like Amazon Web Services. The consequences of such attacks can include a decrease in legitimate traffic, loss of business, and harm to reputation. As stated in the *Encyclopedia of Information Technology* (2007):

> DDoS attacks function by overwhelming devices, services, and networks with bogus internet traffic, rendering them unusable for legitimate users. They are more sophisticated and expansive than traditional DoS attacks, employing thousands or even millions of interconnected devices to achieve their objective. Identifying a DDoS attack can be challenging because the symptoms, such as sluggish network performance or website unavailability, can resemble non-malicious issues.

In Africa, the number of botnet victims detected totaled around 50,000, with an average monthly detection rate of 3,900. There have been numerous notable instances of such DDoS attacks on critical infrastructure in Africa over the past five years (INTERPOL Report, 2021). Therefore, to mitigate the impact of DDoS attacks, organizations can opt for DDoS protection services, devise disaster recovery plans, bolster security measures, and monitor network traffic for anomalous patterns.

8.4.8 Cyberextortion

Cyberextortion refers to the act of compelling a victim to provide funds or personal information through online methods. One prominent tactic of cyberextortion involves installing ransomware on a victim's device or network, which demands payment to restore access. Zuckerman et al. (2011) outlined various methods through which cyber extortion can occur, including ransomware attacks, DDoS attacks, doxing, phone extortion, and website extortion. Ransomware encrypts the victim's files, demanding payment for decryption, while DDoS attacks flood the victim's network with internet traffic, leading to a shutdown until a ransom is paid.

The repercussions of cyber extortion can be severe, resulting in significant financial and reputational harm. Perpetrators often demand high sums, sometimes exceeding six figures, and may threaten to disclose sensitive data if payment is not made. Protective measures against cyber extortion include employing multifactor authentication, providing employee training on phishing, establishing effective firewalls, and implementing protocols for encryption, password complexity, and software updates.

8.5 IMPLICATIONS

> Africa boasts over 500 million internet users, surpassing other regions like North America, South America, and the Middle East. This user volume relative to the population amounts to approximately 38%, with expectations of continued growth due to accelerated digitalization. Key countries leading in internet penetration include Kenya, with 83% of its population online, Nigeria with 60%, and South Africa with 56%.
>
> (Council on Foreign Relations, 2019)

Notably, mobile banking is extensively utilized in these nations, namely Kenya, Nigeria, and South Africa, contributing to Africa's active involvement in digital financial services (Cision, 2020). According to the INTERPOL Report (2021), cybercrime represents a significant future threat, particularly with the proliferation of malicious apps targeting vulnerabilities in mobile devices. Alarmingly, more than 90% of African businesses operate without essential cybersecurity measures in place (CGTN, 2020). An in-depth examination of cybercrime within the African digital landscape, particularly in West Africa, reveals a multifaceted environment influenced by various factors such as technological advancements, economic constraints, and geopolitical dynamics. The prevalence of cybercrime in the region is closely linked to the continent's rapid digital evolution and the vulnerabilities inherent in such progress.

Consequently, African cyber threat assessments underscore the dynamic nature of cyber threats in Africa, highlighting the continent's susceptibility to both cyber-dependent and cyber-enabled crimes. The emergence of cybercriminal organizations, including non-state entities and terrorist groups, presents formidable obstacles to regional cybersecurity endeavors. The COVID-19 pandemic has exacerbated cyber threats, precipitating increased job losses and alterations in working methodologies that have opened new avenues for cybercriminal activities. Africa's digital sphere exhibits significant growth in internet accessibility, particularly via mobile devices, with countries such as Kenya, Nigeria, and South Africa taking the lead in online connectivity. However, despite this advancement, the digital divide persists, particularly in rural areas, underscoring the necessity for enhanced cybersecurity infrastructure and awareness initiatives aimed at safeguarding users against cyber threats. The prevalent absence of cybersecurity protocols in numerous African businesses, with 90% operating without adequate safeguards, leaves the region vulnerable to exploitation by threat actors and substantial financial repercussions.

According to Welance (2023), cybercrime poses an escalating threat to individuals in the digital era, leaving them increasingly susceptible to various detrimental effects such as financial loss, identity theft, emotional distress, and harm to their reputations. Businesses also experience adverse consequences including financial losses, reputational damage, legal ramifications, and the loss of intellectual property. The integrity of an individual reflects the integrity of all as self-protection in a way is an extension of communal protection since an individual is a microcosm of society (Osimen et al., 2023).

Moreover, cybercrime extends its impact to the broader society, causing economic repercussions, raising national security concerns such as social and political instability, facilitating social media manipulation, and disrupting medical services while compromising sensitive patient data. Thus, the activities of cybercriminals affect individuals, the economy, organizations, and society. Cybercrimes targeting individuals include email spoofing, spamming, phishing, cyberstalking, cyber defamation, and voyeurism. Crimes against property encompass credit card skimming, software piracy, domain name disputes, internet time theft, and identity theft. Organizations face cyber threats such as hacking, denial of service attacks, email bombing, Salami attacks, logic bombs, data diddling, and spyware infiltration. Cybercrimes against society involve activities like pornography dissemination, the sale of illegal articles, online gambling, and cyber terrorism (Tubake, 2013).

8.6 SOLUTIONS VIA ARTIFICIAL INTELLIGENCE

AI performs multitask functions (anomaly detection, predictive analytics, fraud detection, and incident response) in dealing with cybercrime in the African digital space. Anomaly detection, also known as outlier detection, identifies unexpected events or observations that significantly differ from the norm. In cybersecurity, it can analyze thousands of data streams to detect changes in access requests, an increase in failed authentications, or unusual traffic patterns that warrant further investigation. Anomaly detection is integrated into many security appliances and services, such as intrusion detection systems, web application firewalls, and API security tools. In banking and finance, it is often used to identify fraud by correlating factors like transaction size, time, location, and spending rate. Examples of AI anomaly detection tools include Anodot, Amazon SageMaker, ELKI, PYOD, Scikit-learn, and Wolfram Mathematica.

Predictive analytics involves using computer and machine learning models to analyze data and make decisions based on various factors and insights. Examples of AI predictive analytics software include Qlik, OneModel, SAS VIVA, Scios, and SAP Analytics Cloud. Fraud detection involves identifying and preventing fraudulent activities within applications, APIs, systems, transactions, and data. Examples of fraud detection software used in financial institutions like banks include Swift GPI, EBA CLEARING Fraud Pattern and Anomaly Detection (FPAD), MasterCard Consumer Fraud Risk, and Salv Bridge. Incident response involves the immediate management of data breaches or cyber-attacks to mitigate the risk of future incidents. This process includes several steps: preparation, detection and analysis, containment, eradication and recovery, and post-incident activities (Cichonski et al., 2012). Examples of AI tools for incident response are Microsoft Security Copilot, Tessian's Complete Cloud Email Security, Cylance, Cybereason, and McAfee MVISION. AI solutions (fraud detection) in tackling cybersecurity challenges in Africa are: Safaricom (Kenya), IBM Research Africa (South Africa), Andela (Nigeria), E-Tranzact (Ghana), and Tigo (Tanzania).

8.7 KNOWLEDGE MANAGEMENT, ARTIFICIAL INTELLIGENCE, AND CYBERSECURITY IN AFRICA DIGITAL SPHERE

Knowledge management is an integral component in the enhancement of new information and communication technologies as well as in the progress and coordination of various systems (Arjoun & Boudabbous, 2024). Knowledge management encompasses the entire process of discovering and creating knowledge, along with its distribution and application (Uto et al., 2024). Onyeaghalaji and Igberaese (2010) argue that proper implementation of knowledge management and capacity building would foster a harmonious environment, facilitate the coordination of ideas and thoughts, and ensure the integration of knowledge, ultimately promoting human development. Knowledge management systems can streamline and automate the knowledge management processes within organizations. They support the discovery, sharing, application, and capture of knowledge throughout the organization (Riswanto & Sensuse, 2021).

Knowledge management systems integrate people, processes, and technology (Dei, 2021). Knowledge management helps in understanding knowledge, whereas AI offers the tools to expand, utilize, and generate knowledge in innovative ways that are beyond our current imagination (Smuts & Borgstein, 2021). African organizations leveraging Artificial Intelligence and Knowledge Management to improving cybersecurity are: African Union's Cybersecurity Initiative, Kenyan Government's National Computer and Cybercrimes Coordination Committee, South African Financial Sector's Cybersecurity Initiatives, Nigerian Cybersecurity and Infrastructure Protection Initiative, and Pan-African Organization for Cybersecurity (PACO). Ways by which knowledge management plays a vital role in disseminating AI-driven cybersecurity tools across various sectors in Africa are: Collaboration and networking platforms; AI-enhanced knowledge discovery; training and capacity building; integration with existing systems (security operations centers); and monitoring and evaluation. Examples of AI-driven tools for knowledge management in Africa are: Jumia (manages and analyzes vast amounts of customer data), MEST Africa (manages educational content and resources), LexisNexis South Africa (manages legal documents and ensures compliance), iCow (uses machine learning to analyze agricultural data), Vezeeta (manages and analyzes patient data and medical records) and Rwanda Development Board (RDB) (translates documents and information into Kinyarwanda and other local languages).

8.8 CONCLUSION

The essence of digital technology is to better man's welfare, society, and to put an end to communication barriers between individuals, groups, nations, and continents. Unfortunately, some irresponsible people see digital technology as an avenue to make money dishonestly and illegally. The African digital space is threatened by varied forms of cybercrimes and their negative impact on African nations is so dense impeding societal growth and progress. Cybercriminals, irrespective of the category, have advanced in their mode of operations, getting more sophisticated as the day goes by. Cybercrime is a reality as it presents a multifaceted threat to national security, economic prosperity, and the digital landscape in Africa. It not only impacts government revenue streams but also deters entrepreneurial endeavors on the internet, resulting in significant financial losses amounting to billions of dollars. The surge in internet usage and digital innovations has amplified and transformed various national security challenges and risks, including espionage, sabotage of critical infrastructure, and organized crime activities. As such, it becomes imperative for African nations to bolster their cybersecurity defenses with the aid of advanced AI. From these findings, the following recommendations are made in combatting cybercrimes in African digital space:

i. The strategy to combat cybercrime in Africa should involve a multifaceted approach, encompassing initiatives such as raising awareness among traditional communities, fortifying policy frameworks, and bolstering technological capabilities for cyber defense. Collaborative efforts among regional entities like INTERPOL, the African Union, AFRIPOL, ECOWAS, ICPC, and EFCC are crucial for maximizing coordination and capacity-building endeavors aimed at effectively addressing cyber threats.

ii. Safeguarding the African digital space and mitigating the escalating cyber threats necessitates a comprehensive approach in Africa, which integrates technological advancements, policy frameworks, public awareness campaigns, and regional cooperation.

iii. Utilizing cybercrime investigators and investigative tools is essential for combating cyber threats effectively. These tools include Digital Forensics Software for recovering deleted files and analyzing metadata, Network Analysis Tools for monitoring network traffic and identifying suspicious activity, Malware Analysis Tools for dissecting and understanding malware behavior, Password Recovery Tools for decrypting passwords from encrypted files, databases, or digital evidence sources, and Social Media Analysis Tools for tracking suspects' activities and gathering evidence from social media platforms.

iv. The use of digital forensics, which involves the collection, preservation, and analysis of digital evidence, is paramount in combating cybercrime. This process encompasses tasks such as recovering deleted files, scrutinizing metadata, and examining network traffic logs, facilitated by software tools like EnCase, FTK, and Autopsy.

v. Strengthening cybersecurity measures, raising public awareness, and formulating robust policies are vital components of the strategy to combat cyber threats effectively in Africa. This includes joint and collaborative efforts with stakeholders in the cyber industry, implementation of national cybersecurity policies through national strategies, enactment of legislation against cybercrime, and establishment of national regulatory authorities.

vi. African government should establish more cyber incident response centers and fortify them accordingly.

REFERENCES

Abasilim, U. D., Gberevbie, D.E., & Osibanjo, O. A. (2019). Leadership styles and employees' commitment: Empirical evidence from Nigeria. *Sage Open 9*(3), 10.1177/2158244019866287

Aftab, P. (2006). Pobrano Z. http://www.wiredsafety.net

Ajijola, A.-H., & Allen, N. D. F. (2022). African lessons in cyber strategy. *African Center for Strategic Studies*. https://africacenter.org.spotlight.african-lessons-in-cyber-strategy/

Arjoun, A., & Boudabbous, S. (2024). Knowledge management: A challenge for the company. *International Journal of Business and Management Review*, *12*(1), 81–95.

Atlantic. (2007). *Encyclopaedia of Information Technology*. Atlantic Publishers & Distributors.

Borges, E. (2024). Cyber criminals: Tactics, impacts, and defense strategies. https://www.recordedfuture.com/threat-intelligence-101/threat-actors/cybercriminals

Campell, M. (2015). Cyberbullying: An old problem in a new guide. *Australian Journal of Guidance and Counselling 15*, 68–76.

CGTN. (2020). Rights group launches tool to stem cybercrime in Africa. https://news.cgtn.com/news/2020-10-27/Unveiling-the-cost-of-cybercrime-in-AfricaUVhmu1PJeM/index.html

Chaubey, R. K. (2012). *An Introduction to Cyber Crime and Cyber Law*. Kamal Law House.

Cichonski, P., Millar, T., Grance, T., & Scarfone, K. (2012). *Computer Security Incident Handling Guide*. Special publication 800-61, Revision 2, National Institute of Standards and Technology, Department of Commerce, US. S

Cision. (2020). Africa leads world in digital financial services deployments with prepaid cards an important part of mix, says Axiom prepaid holdings reps. https://www.prweb.com/releases/Africa_leads_world_in_digital_financial_services_deployments_with_prepaid:cards_an_important_part_of_mix_says_axiom_prepaid:holdings_reps/prweb17214821.htm

Council on Foreign Relations. (2019). Last month, over half-a-billion Africans accessed the internet. https://www.cfr.org/blog/last-month-over-half-billion-Africans-accessed-internet

Cyber Talent. (2023). Cybercrime investigation tools and techniques you must know! https://cybertalents.com/blog/cyber-crime-investigation

Daudu, B. O., Osimen, G. U., & Shuaibu, K. (2023). Cyberbullying and social media: Implications for African digital space. In Mohamed Lahby, Al-Sakib Khan Pathan & Yassine Maleh (Eds.), *Combatting cyberbullying in digital age with artificial intelligence* (pp. 243–253). Chapman and Hall/CRC Press.

Dei, De-G. J. (2021). Perspectives of knowledge management systems implementation. *Library Philosophy and Practice* (e-journal). S. https://digitalcommons.unl.edu/libphilprac/5610

Economic Commission for Africa. (2022). ECA launches the guideline for a Model Law on Cyber security during the 17th IGF. https://www.uneca.org/stories/eca-launches-the-guideline-for-a-model-law-on-cybersecurity-during-the-17th-igf

FBI. (2021). *Internet Crime Report*. Internet Crime Complaint Center.

Ganacharya, T. (2020). Protecting against coronavirus-themed phishing attacks. *Microsoft Security blog*. https://www.microsoft.com/en-us/security/blog/2020/03/20/protecting-against-coronavirus-themed-phishing-attacks/

Gberevbie, D., Oyeyemi, A., & Nchekwube, E.-O. (2014). The challenges of good governance, accountability of governmental agencies and development in Nigeria. *Acta Universitatis Danubius*, *6*, 80–97.

Halder, D., & Jaishankar, K. (2011). Cyber gender harassment and secondary victimization: A comparative analysis of the United States, the UK, and India, *Victims & Offenders*, 6(4), 386–398. http://doi.org/10.1080/15564886.2011.607402

Hunter, et al. (2007). Perceptions and correlates of peer victimization and bullying. *British Journal of Educational Psychology* *77*(4), 797–810.

Hussain, A. (2004). *What Is Identity Theft? Types and Examples*. Investopedia.

INTERPOL Report. (2021). *African Cyberthreats Assessment Report: Interpol's key insights into Cybercrime in Africa*. INTERPOL Global Complex for Innovation.

King, L. (2006). No hiding from online bullies. http://www.news-leader.com/apps/pbcs.doll/articles?Date=20060815

Kizza, J. M. (2014). *Computer network security and ethics* 4th edition. McFarland & Company, Inc., Publishers.

Kundi, G. M., & Akhtar, R. (2014). Digital revolution, cyber-crimes and cyber legislation: A challenge to governments in developing countries. *Journal of Information Engineering and Applications*, 4(4), 61–70.

Lehman, J., & Shirelle, P. (2005). *West's Encyclopedia of American Law* Vol. 3, 2nd edition. Thomson/Gale.

Mativat, F., & Tremblay, P. (1997). Counter-feiting credit cards. *British Journal of Criminology*, 37(2), 165–183.

Mishra, R. (2023). Cybercrime and safety. https://www.studocu.com/in/document/swami-ramanand-teerth-marathwada-university/llb-3-year/cybercrimeandsafety-good/82727988

Onyeaghalaji, M. N., & Igberaese, D. E. (2010). Knowledge management and capacity building for sustainable development in Africa. *Journal of Sustainable Development in Africa*, 12(2), 268–283.

Osimen, G. U., Daudu, B. O., & Awogu-Maduagwu, E. A. (2023). Nigerianness and misrepresentations of names and signatures: Should the Colonized adopt the english-language naming conventions? *Journal of the International Society for the Study of Vernacular Settlements* (ISVS e-journal), 10(11), 289–303. https://doi.org/10.61275/ISVSej-2023-10-11-19

Positive Technologies. (2023). Cybersecurity threatscape of African countries 2022-2023. https://ptsecurity.com/ww-en/analytics.africa-cybersecurity-threatscape-2022-2023/

Riswanto, & Sensuse, D. I. (2021). Knowledge management systems development and implementation: A systematic literature review. *IOP Conf. Series: Earth and Environmental Sciensce*, 704. 012015. https://doi.org/10.1088/1755-1315/704/1/012015

Roberts, T. (2021). Digital rights in closing civic space: Lessons from ten African countries. https://www.ids.ac.uk/opinions/digital-rights-and-closing-civic-space-lessons-from-ten-african-countries

Shariff, S. (2009). *Confronting Cyberbullying: What Schools Need to Know to Control Misconduct and Avoid Legal Consequences*. Cambridge University Press.

Smuts, H., & Borgstein, C. (2021). Artificial intelligence (AI) and knowledge management (KM): two sides of the same coin? https://realkm.com/2021/11/14/artificial-intelligence-ai-and-knowledge-management-km-two-sides-of-the-same-coin/

Tahiru, A. (2018). Cybersecurity in Africa: The Threats and Challenges. *Cyberpolitik Journal*, 3(5), 91–104. www.cyberpolitikjournal.org

Toulas, B. (2022). OPERA1ER hackers steal over $11 million from banks and telcos. https://www.bleepingcomputer.com/news/security/opera1er-hacj=kers-steal-over-11-million-from-banks-and-telecos/

Tubake, M. (2013). Cyber crimes: An overview. *Online International Interdisciplinary Research Journal*, 3(2), 129–139.

Uto, S.C., Uwa, K.L., & Akpan, A. (2024). Knowledge management and competitive advantage in selected manufacturing firms in Akwa Ibom state. *International Journal of Business and Management Review*, *12*(1), 1–20. doi: https://doi.org/10.37745/ijbmr.2013/vol12n1120

Viswanathan, S. T. (2001). *The Indian Cyber Laws With Cyber Glossary*. Bharat Law House.

Welance. (2023). Impact of Cybercrime on individuals, businesses, and society. https://www.linkedin.com/pulse/impact-cybercrime-individuals-businesses-society-join-welance

Zola, A. (2023). Digital signal processing (DSP). *Technology Journalist*, 5. https://www.techtarget.com/whatis/definition/digital-signal-processing-DSP

Zuckerman et al. (2011). *Distributed Denial of Service Attacks Against Independent Media and Human Rights Sites*. The Berkman Center for Internet & Society at Harvard University.

9 Enhancing Knowledge Management Integrity through Fake News Detection

A Hybrid LSTM-XGBoost Approach for Cybersecurity

M. V. Sujan Kumar, Ganesh Khekare, and Anurup Sankriti

9.1 INTRODUCTION

As society navigates through a complex web of digital media, the credibility of information is at stake, necessitating robust measures to distinguish between authentic and deceptive content. This chapter addresses the pressing concern of fake news within the realm of cybersecurity, leveraging a dataset that encompasses both textual content and metadata, providing a comprehensive foundation for training and testing models. To enhance the precision and efficiency of fake news detection within the cybersecurity domain, we adopt a hybrid approach that synergizes the strengths of short-term memory (LSTM) neural networks with a customized version of the XGBoost gradient boosting algorithm. By amalgamating the predictive capabilities of XGBoost with LSTM's adeptness in discerning intricate patterns in the dataset, we strive to construct a more holistic and efficient system for identifying cyber threats and deceptive information. Through the exploration of these data, our objective is to not only contribute to the existing knowledge on fake news detection but also to delve into the potential of the LSTM-XGBoost hybrid model within the cybersecurity context. This chapter endeavors to elucidate the pivotal features and patterns that differentiate genuine cybersecurity information from deceptive content, laying the groundwork for the development of advanced and reliable media authentication technologies tailored to the intricacies of the cyber threat landscape. As cybersecurity is constantly progressing, the proper storage of data has always been a significant element for ensuring security from threats in the context of developing knowledge management. The issue of fake news, a growing problem in the organization, impacts knowledge management systems due to increased misinformation,

DOI: 10.1201/9781003498094-9

TABLE 9.1
Keywords Used in the Chapter and Their Abbreviation

Keyword	Abbreviation
Long Short-Term Memory	LSTM
Extreme Gradient Boosting	XGBoost
Support Vector Machine	SVM
K-Nearest Neighbors	KNN
Area Under the Curve	AUC
Receiver Operating Characteristic	ROC
Fake News Detection Network	FNDNet

social engineering, and manipulations of key decision-making processes. Therefore, by incorporating machine learning algorithms such as Long Short-Term Memory (LSTM) and eXtreme Gradient Boosting (XGBoost) in the knowledge management frameworks, organizations can improve their capacity to address the flow of fake news and weaken the knowledge management risks. The keywords used in the chapter and their abbreviation are shown in Table 9.1.

9.2 LITERATURE REVIEW

9.2.1 FNDNet for Fake News Detection

FNDNet, a deep neural network devised by Cuevas et al., is specifically crafted for detecting fake news. Unlike conventional techniques, FNDNet employs a deep CNN architecture, which enhances its capability to identify distinctive features. Demonstrating an impressive 98.36% accuracy rate in data analysis, this methodology stands as a pivotal advancement in fake news detection. The model not only bolsters trust and confidence in communication media but also lays the groundwork for further interdisciplinary research, integration, and translational studies. Its efficacy in augmenting the reliability of news sources underscores its significance in combating misinformation.

The deep convolutional neural network structure of FNDNet facilitates more efficient learning of text feature representations, thereby enhancing its capacity to discern fake news. Characterized by multiple parallel convolutional layers with varying kernel sizes, this architecture adeptly extracts diverse n-gram features, enriching the coverage of the language model. Through strategic layer packing to minimize size, the network culminates in a configuration featuring a single flat layer, two dense layers, 0.2 dropouts, and ReLU activation. Trained on word embeddings, FNDNet's remarkable 98.36% accuracy signifies a substantial leap forward from prior machine learning and deep learning methodologies. Notably, its automatic feature engineering mitigates challenges associated with limited hand-generated feature data, paving the way for future endeavors encompassing multimodal fake news detection, integration of social context, and real-world application. Fake news identification method is much needed as large amounts of data in the form of blogs are circulated over the internet.

9.2.2 Graph Neural Networks for Fake Post Detection

In their paper, Izotova and Lavrova demonstrate the effectiveness of using graph neural networks to model the spread of misinformation and detect misinformation on social media. They created a graph model where nodes represent posts, resources, and notes, while edges represent non-fiction spread. After examining different network architectures, they settled on a model that describes the structure of ads in the user network, including features such as the number of ads, and user metadata (identification status and followers, etc.). Sort posts into categories. Their key insight is that untrustworthy users are more likely to post fake content and engage with other dubious accounts. By editing the interactions in the image and feeding them into the neural network, its models can reliably distinguish real posts from fake ones. Tests on two datasets demonstrate the ability of neural networks to model complexity and detect false content, demonstrating high precision and recall. The custom graphical model allows them to show the relationship between inaccurate data in detail and improves classification accuracy compared to using reports and data alone. Their work shows the promise of using deep learning on datasets to combat the spread of fake news on social media.

9.2.3 Instagram Fake News Detection System

This paper proposes an automated fake news detection system tailored for Instagram. The authors generate novel content and context features based on Instagram's unique characteristics, such as average comment length, engagement rate, and number of fake followers. These features are combined with common attributes from literature like the number of likes and hashtags. Several machine learning models including SVM, logistic regression, Naive Bayes, random forest, and KNN are applied to a manually labeled Instagram dataset collected by the authors. The results demonstrate that the proposed features improve fake news classification accuracy across all the models. Specifically, the KNN algorithm achieves a 99% F1 score using the augmented feature set. The most discriminative features are found to be post count, URL in bio, average comments, engagement rate, average likes, and fake followers. This indicates that user behavior and credibility cues are more predictive compared to textual content for Instagram fake news detection. Overall, the study shows how creating social media-specific feature sets can boost the performance of classical Machine Learning techniques. The models and Instagram dataset developed can serve as a strong baseline for future work on multimodal fake news classification leveraging images and text.

9.2.4 Psychology-Informed Fake News Detection

Zhou et al. propose a novel fake news detection model that focuses on analyzing news content rather than propagation patterns. Their key innovation is introducing theories from social psychology like the Undeutsch hypothesis and reality monitoring into the feature engineering process. By representing news articles across lexical, syntactic, semantic, and discourse levels guided by these theories, their interpretable feature set enables early fake news detection and analysis of linguistic patterns (Daneshvar et al., 2024). Experiments on two benchmark datasets demonstrate that their model outperforms state-of-the-art methods for early fake news detection relying solely on content. The authors recommend further improvements by incorporating user metadata, additional NLP techniques like named entity recognition, and more attribution analysis between fake news and other forms of deception (Raja et al., 2024). Their interdisciplinary theory-driven approach highlights explainable differentiators between real and fake news while enabling early detection before propagation. This direction of interpretable models guided by the social sciences holds promise for further progress on the important challenge of fake news detection (Khekare and Verma, 2021).

9.2.5 Albanian Language Fake News Detection

This research presents the first-ever fake news corpus and benchmark for the low-resourced Albanian language. The authors compile a dataset of 2,000 true and 2,000 fake news articles, carefully aligning them by topic similarity. They extract 267 linguistic features spanning lexical, syntactic, psycholinguistic, deception, and emotional dimensions. These engineered features are tested on classifiers like SVM and XGBoost using full-length and truncated versions of the true articles to mitigate possible length bias. Experiments demonstrate exceptional results, with top methods catching over 90% of fake news on truncated texts. The features prove highly discriminative despite Albanian's limited NLP resources. While article length does not significantly advantage classifiers as hypothesized, the authors recommend future work to enrich the corpus with more metadata and focus on improving the representation of Albanian linguistic structures. This study pioneers a fake news detection approach for low-resource languages, leveraging deep linguistic feature engineering to uncover deception signals. It provides a benchmark and groundwork to advance Albanian NLP and combat online misinformation. The DHCF model uses two communication theories to represent education and information distribution (Craig and Gainous, 2024). The model is more efficient than existing systems and shows improvements in stability and accuracy. The report recommends adding a resolution to achieve better interpretation, increase computational efficiency, and extend the model to accommodate multimode devices (Choudhary et al., 2024).

9.2.6 Deep Learning-Based Fake News Detection

The article proposes a deep learning-based fake news detection system that utilizes both convolutional and recurrent neural networks. Past works have applied a variety of machine learning techniques like random forests, support vector machines, and neural networks for fake news classification (Cavalcante et al., 2024; Choudhary and Arora, 2024). Recently, the use of deep learning has become popular given its ability to learn high-level feature representations. The proposed cascaded CNN model extracts n-gram level text features, while the LSTM captures long-term sequential dependencies in news articles. Similar concatenated CNN-RNN architectures have shown promising results on fake news benchmarks. A key novelty of this work is the comparison between multiple deep network variants on the same news corpus. The experiments reveal that a 3-branch concatenated CNN achieves the best accuracy of 99.6%, outperforming LSTM and GRU models. The multi-dataset evaluation demonstrates the robustness of deep learning against rule-based and classical ML classifiers. Overall, the work provides valuable insights into optimal neural designs and the feasibility of deep learning for real-time fake news detection.

9.2.7 Clickbait and Fake News Detection

Fake news and misinformation spreading on social media have become a major concern in recent years. Researchers have proposed various methods to detect fake news using machine learning and natural language processing techniques. Goyal et al. (2024) provided a comprehensive survey of fake news detection methods. Zhou et al. (2020) used stance detection to facilitate fake news detection, achieving over 80% accuracy on a fake news dataset. Sedik et al. (2022) proposed a system to detect and prevent clickbait in online news. Bondielli and Marcelloni (2019) discussed methods for automatic deception detection to find fake news. Alwaro focused on lightweight fake news detection using headline features. They extracted attributes like capital words, punctuation marks, and user engagement metrics to train classifiers. Their method achieved 99.4% precision using a logistic regression classifier, demonstrating excellent performance for identifying potential clickbait stories. As public fake news datasets become available, their approach could be expanded to further improve generalization. Overall, machine learning provides promising capabilities for automatic fake news detection, which will grow increasingly important for news consumers and social media platforms.

9.2.8 Automated Detection of Fake Job Posts

Dutta and Bandyopadhyay proposed an automated machine-learning approach to detect fake job posts online and protect job seekers from employment scams. They developed classification models using both single classifiers (Naive Bayes, MLP, KNN, Decision Tree) and ensemble techniques (Random Forest, AdaBoost, Gradient Boosting). After evaluating on a dataset of over 17,000 job posts, the Random Forest ensemble classifier achieved the best performance with 98.27% accuracy (Canhasi et al., 2022). The key innovation of this research is demonstrating how ensemble learning methods that combine multiple decision tree models can outperform individual machine learning classifiers for identifying fraudulent job advertisements (Farhangian et al., 2024). By comparing metrics like accuracy, F1 score, Cohen's Kappa, and mean squared error, the authors provide evidence that ensemble techniques like Random Forest are well-suited for the complex pattern recognition task of fake job posting detection. This has important implications for developing automated systems to detect employment scams and prevent the exploitation of job seekers in the online space (Ganesh Khekare et al., 2023).

9.2.9 Multi-Pronged Approach to Fake News Detection

FNDNet is a deep neural network specifically designed to detect fake news (Khekare and Midhunchakkravarthy, 2023). The algorithms use a range of conversion techniques to recognize unique features and determine the authenticity of news; hence excelling at detecting fake news (Guo et al., 2023). The authors point to potential cases for anti-fake news solutions by providing a pathway for the next steps in research, including testing of materials in larger formats, integration, search translation, infrastructure development, and protection against attacks (Wang et al., 2023). These studies demonstrate a variety of strategies used to solve the problem of identifying fake news, from content-based methods and specific studies to collaborative learning (Izotova and Lavrova, 2021), abstract architectures (Koru and Uluyol, 2024), and graph neural networks (Khekare et al., 2022). Collectively, the findings provide a foundation for future research and highlight the need for robust, comprehensive (Maity et al., 2024), and meaningful methods to combat this widespread problem. The summary of the approaches used by the previous researchers for Fake News Detection is shown in Table 9.2.

9.3 METHODOLOGY

Data Acquisition: The initial step involves loading the pre-processed dataset ('cleaned_train_data.csv') containing text articles and their corresponding binary labels (0 for real, 1 for fake).

Train-Validation Split: The dataset is divided into training and validation sets using an 80–20 split through the train_test_split function from sklearn. Model_selection.

Text Tokenization and Padding: The text data in the 'text' column is tokenized using the Tokenizer from Keras, which converts the text into sequences of integers based on word frequency. Vocabulary Size and Padding: Tokenized sequences are then padded to ensure uniform length using pad sequences. The maximum sequence length ('male) is set to 300, accommodating longer texts by padding or truncating where necessary.

LSTM Model Creation

Architecture Setup: A Sequential model in Keras is constructed for the LSTM network. Embedding Layer: An Embedding layer is added to create word embeddings based on the tokenized sequences with a specified vocabulary size and input length (Khekare et al., 2024).

Bidirectional LSTM Layers: Two Bidirectional LSTM layers are stacked for capturing context from both directions, employing 256 and 128 units, respectively, with dropout for regularization.

Dense Layers: Dense layers with ReLU and Sigmoid activations are added for higher-level feature extraction and binary classification.

LSTM Model Training

Model Compilation: The LSTM model is compiled using binary cross-entropy loss and the Adam optimizer.

TABLE 9.2
The Summary of the Approaches Used by the Previous Researchers for Fake News Detection

Article	Approach Used	Pros	Cons
Maningo and Hlaing (2024)	Logistic Regression Random Forest Bayes Net	The paper proposes a practical solution using a browser extension tool to detect and filter out potential fake news/clickbait sites based on features like the synthetic structure of titles, the use of punctuation, and high bounce rate. This could be useful for readers to avoid misleading sites.	The dataset collection process is not clear. More diverse data would have established robustness. No comparison with other state-of-the-art fake news detection methods.
Manzoor et al. (2019)	GNN, GCN	GNNs are effective at learning from graph structure and node features. Achieves good performance of 0.9688	Computation is the intensive cause of GNNs as compared to other models. Constructing the heterogeneous graph requires access to additional data like social networks which may not always be available.
Aldwairi and Alwahedi (2018)	Decision Tree, KNN, Ada Boost	Good size of dataset close to 18k job posts. Systematically evaluates multiple models using relevant metrics.	Limited technical details for pre-processing, and feature extraction steps. The splitting of the dataset into train test split is not mentioned.
Mehravaran and Shamsinejadbabaki (2023)	K-nearest Neighbors Classifier Gradient Boosting Multi-Layer Perceptron Classifier	Compares several machine learning classifiers for the challenge of detecting bogus job posts. Demonstrates that ensemble classifiers such as Random Forest outperform solo classifiers. Achieves very excellent accuracy of 98.27% using the Random Forest classifier.	There is no information available about the specific features employed or the feature engineering procedures. Does not assess performance on unobserved real-world data, simply the training-test split of the dataset. No comparison with other contemporary approaches in the literature for benchmarking.

(Continued)

TABLE 9.2
(Continued)

Article	Approach Used	Pros	Cons
Chang et al. (2024)	XGBoost, SVM, Random Forest, BERT, Word2Vec (learning Techniques)	XGBoost, SVM, and random forests gave the best performance overall. Engineered features like lexical, syntactic, lying-detection, and psycholinguistic performed well with 85–90% F1 measure. Representation learning techniques like Word2Vec and BERT also gave good accuracy of 90%+.	Performance dropped when texts were truncated to remove length bias. Dialect classifiers did not perform as well as expected. Advanced language models could not handle noisy texts well, so did not improve much over bag-of-words. Performance dropped significantly when news sources were unreliable, indicating writing style plays a big role.
Kaliyar et al. (2020)	CNN (Convolutional Neural Network) C-CNN (Concatenated CNN) CNN-LSTM (CNN combined with Long Short-Term Memory network) GRU (Gated Recurrent Unit network)	Achieved high accuracy (98.6% for CNN, 99.9% for C-CNN), faster training time. Performed better for more complex relationship detection tasks like analyzing the relationship between article headlines and body text (96% accuracy for CNN-LSTM). More capable of sequence-based learning.	Not as effective for more complex relationship detection tasks. Slower training time compared to CNN/C-CNN models. Lower accuracy on simpler fake news detection tasks.
Santhiya et al. (2023)	SVM Random Forest k-Nearest Neighbors	Combines content and context features for better accuracy. Proposes new features tailored to Instagram like engagement rate and number of fake followers. Tests on a new Instagram fake news dataset created by the authors. Experiments with multiple machine learning models to compare performance.	The Instagram dataset created is still relatively small with 5028 posts. More data could further validate the results. Deep learning techniques were not explored; only shallow machine learning models were used.

Fang et al. (2024)	Multimodal	Enables early fake news detection before propagation by solely relying on content. Interpretable features based on theories can help understand fake news patterns. Generally, it achieves high accuracy around 88% on benchmark datasets. Resilient to changes in data size and news distribution.	Handcrafting features can be labor-intensive. Heavily dependent on psychological theories which may not fully generalize. Limited exploration of images and rhetorical relationships Unable to directly measure some attributes like readability.
Yu et al. (2019)	Dual disentangled-view hierarchical contrastive learning (DHCF) model	Captures both structural (via propagation graph) and temporal (via propagation sequence) features of news propagation to model news spread effectively. Uses disentangled representation learning through Disen-GraphEnc and Disen-SeqEnc to infer underlying intentions behind interactions between posts. This enhances model expressiveness. Employs hierarchical contrastive learning framework with intra- and inter-view contrasts to handle limited labeled data and improve robustness against noise/unreliability. Significantly outperforms state-of-the-art methods on benchmark datasets, especially with limited training data and under noise attacks.	Complex model architecture with multiple components makes implementation and hyperparameter tuning more difficult. Contrastive learning and disentangled representation add to computational overhead for training. Needs further testing on more diverse and larger-scale datasets to establish the generalizability of performance gains. Lacks interpretability compared to some simpler supervised models.

(*Continued*)

TABLE 9.2
(Continued)

Article	Approach Used	Pros	Cons
Zhang et al. (2024)	FNDNet—A deep convolutional neural network for fake news detection	Achieved a very high accuracy of 98.36% on the fake news dataset, outperforming previous state-of-the-art methods. Learns discriminative features automatically through multiple hidden layers instead of relying on hand-crafted features. Uses multiple parallel convolutional layers with different kernel sizes to capture more contextual information. Deeper architecture with more convolution and pooling layers makes it more effective for text classification.	Computationally more expensive to train compared to simpler machine learning models. Requires significant hyperparameter tuning to find the optimal CNN architecture. Still evaluated only on a binary fake/real news classification task, a more complex multi-class problem remains open.

Model Fitting: Training the LSTM model on the tokenized and padded text sequences is executed for 5 epochs with a batch size of 64, validating the validation set during training.

Feature Extraction With LSTM

Prediction on Training and Validation Sets: The trained LSTM model is used to generate features from the text sequences for both the training and validation datasets.

XGBoost Model Integration

XGBoost Model Initialization: An XGBoost Classifier model is initialized.
Model Training: The XGBoost model is trained using the features extracted from the LSTM model on the training set.

Prediction and Performance Evaluation

Prediction with XGBoost: The XGBoost model predicts labels on the validation set features obtained from the LSTM model.
Performance Evaluation: The accuracy score is computed using the accuracy score from sklearn. metrics to assess the performance of the combined LSTM-XGBoost model.

Confusion Matrix Visualization

Confusion Matrix Calculation: A confusion matrix is computed using a confusion matrix from sklearn. metrics to visualize the model's classification performance.
Plotting Confusion Matrix: The confusion matrix is plotted as a heatmap using Matplotlib and Seaborn libraries to display predicted versus actual labels.

9.4 ARCHITECTURE DIAGRAM

The proposed fake news detection framework is shown in Figure 9.1. Our proposed method uses a methodological strategy to solve the problem of false information detection. Starting with painstaking data collection and preparation, we tokenize, numerically encode, and pad text for training and testing. The core of our model is a bidirectional LSTM neural network that starts with an embedding layer and is then enhanced by a second bidirectional LSTM layer for contextual understanding We also include the XGBoost model, which uses gradient-boosted decision trees role. Our approach enables thorough testing with measures such as accuracy and confusion matrix.

Machiner, the LSTM and XGBoost models offer a powerful solution, where modern machine learning techniques can be used to identify textual patterns indicative of false reports, which can be interpreted through visualization. The LSTM-XGBoost hybrid model for detecting fake news shows its significance as a sub-system integrated into a large cybersecurity knowledge management system. Hence, by providing an

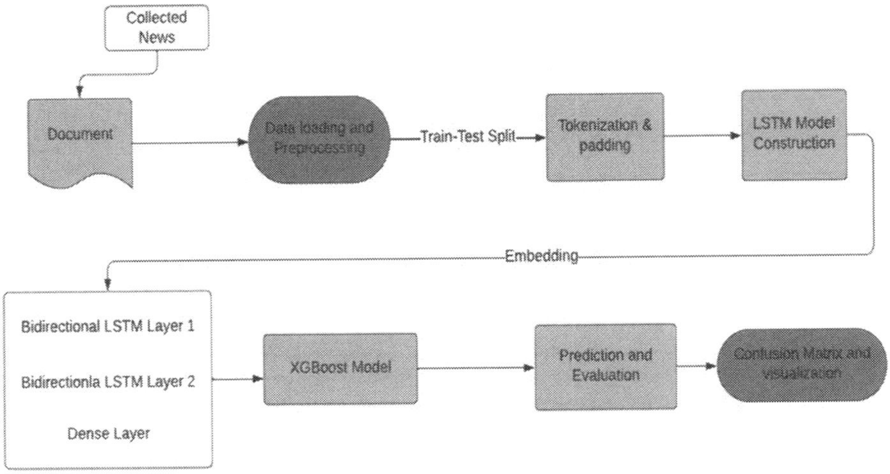

FIGURE 9.1 The proposed fake news detection framework.

accurate fake news identification method, this system supports the enhancement of the reliable knowledge base that security professionals may use. It also contributes to preventing more leaks of information in an organization and assists in enhancing probabilities of smarter threat intelligence and risk assessment services. Moreover, fake news detection helps to enrich the knowledge management system with insights into the specifics of the potential threats and impacts to be expected from the alarming signals as news or updates to enhance the organization's capacity to evaluate the credibility of the identified information.

9.5 DATASET SPECIFICATIONS

The dataset is divided into three primary files: train.csv, test.csv, and submit.csv. Each of these files has several columns that provide details about news items and the labels that match them.

Train.csv

Purpose: Used to train machine learning models.
Column:
id: A unique identifier for each message.
Title: Name of the article.
author: Details of the author of the article.
Text: Content or text.
tag: Binary map showing the accuracy of an article (0 real news, 1 fake news).
Usage: There is a significant amount of data sets, and it is used as a data classification model.

Test.csv

Purpose: Stored to test sample performance.
Column:
id: Unique identifier corresponding to the text in the test text.
Title: Title of the news. Author: Author of the article.
Text: The content or body of the article.
author: Details of the author of the article.
Use: Represents a set of retention tests used to evaluate the effectiveness of the training model. Similar features exist for binary maps for evaluation as well as for the training process.

Submit.csv

Purpose: Used for model estimation and evaluation.
Column:
id: Unique identifier corresponding to the media shipped in the environment.
label: The generated predicted binary classification label is evaluated by the learning model.
Purpose: A hypothesis is a test to predict a model. The Label column contains the predicted binary distribution labels from the training model for evaluation.

9.6 FEATURES AND LABELS

For training and testing models, the datasets (train.csv and test.csv) include columns with textual and categorical characteristics like "title," "author," "text," and "label." Each news article's authenticity categorization is indicated by the values in the "label" column (0 for legitimate news, 1 for false news). The fake news data set description is shown in Table 9.3.

9.7 RESULT AND DISCUSSION

Multivariate Naive Bayes classification achieved 71.48% accuracy in true and false reports. The specificity of 99.63% confirms the ability of the model to reduce false positives. However, with a recall rate of 33.90%, it is difficult for the model to capture all true positives. The F1 score of 50.59% strikes a balance between accuracy

TABLE 9.3
Fake News Dataset Description

Data	Rows	Columns
Train	20,800	5
Test	5200	4
Submit	5200	2

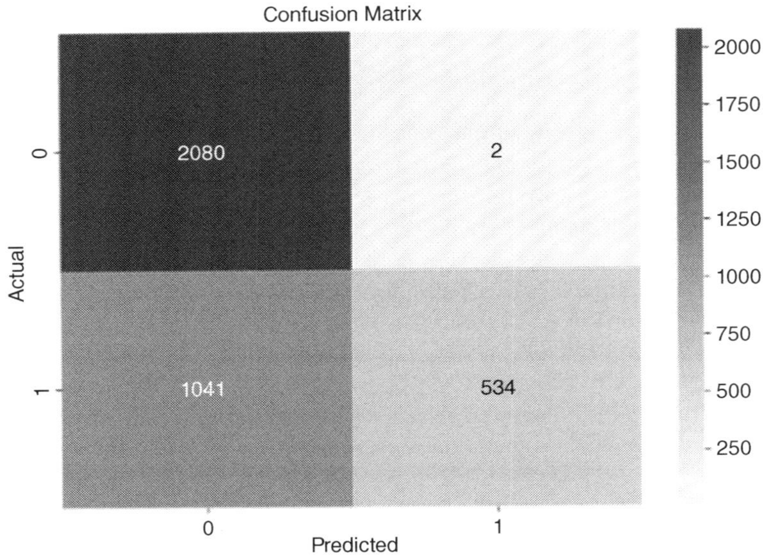

FIGURE 9.2 Naive Bayes confusion matrix with actual and predicted values.

and recall, resulting in a comprehensive performance analysis. The confusion matrix breaks down the model's predictions, showing 2080 true negatives, 534 true positives, 2 false positives, and 1041 false negatives. These metrics summarize the strengths of the Naive Bayes model and potential areas for improvement in false positive detection. The Naive Bayes Confusion Matrix with actual and predicted values is shown in Figure 9.2.

With an accuracy of 90.7%, the Random Forest Classifier effectively distinguishes between genuine and fraudulent information. With an accuracy of 96.83%, the model reduced false positives, indicating reliability. With a recall rate of 81.46%, the model correctly identifies several positive features. The F1 score of 88.48% strikes a balance between precision and recall, indicating the overall performance of the model. The confusion matrix shows 2,040 true positives, 1,283 true positives, 42 false positives, and 292 false negatives. These metrics highlight the ability of random forest classifiers to detect spurious information, with emphasis on strengths and areas for improvement. The Random Forest Classifier Confusion Matrix is shown in Figure 9.3.

The linear regression model showed exceptional prediction, with a mean squared error (MSE) of 2.68e–08. This small accuracy highlights the accuracy of the model in capturing the complex relationships in the data. The MSE measure not only demonstrates the effectiveness of the linear regression method but also establishes it as a powerful tool for regression problems. The ability of the model to reduce prediction errors enhances its robust performance, emphasizing its suitability for applications requiring prediction accuracy.

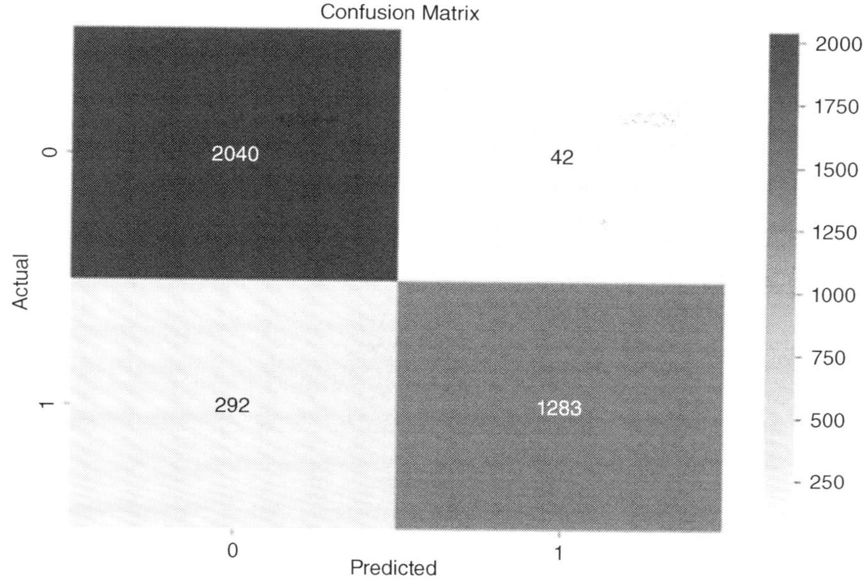

FIGURE 9.3 Random forest classifier confusion matrix.

91.61% accuracy of the CNN model shows better performance in false case detection. This model was trained over 10 epochs and gradually improved, reducing the loss to 0.3354. In terms of accuracy, the model scored 90.49%, resulting in a few false positives. With a recall rate of 89.97%, CNN found very good accuracy. The 90.23% F1 score strikes a good balance between accuracy and recall. The confusion matrix shows 1933 true positives, 1417 true positives, 149 false positives, and 158 false positives. These features reveal the robustness of CNN and provide useful insights into its classification skills for detecting fake news. The CNN Model Confusion Matrix is shown in Figure 9.4.

Five different algorithms were used for comprehensive analysis to detect fake news. Among these models, the LSTM + XGBoost hybrid model emerged as the best model with an accuracy of 94.04%. The random forest classifier also performed very well, with an accuracy of approximately 90.81%. However, the accuracy of the Naive Bayes model lagged by 71%, indicating limitations in capturing relationships in the data. At the same time, the linear regression showed a very low error (2.68e–08), indicating a good predictive power for the regression function. The Convolutional Neural Network (CNN) performed very well, with an accuracy of approximately 91.61%, demonstrating the ability to learn hierarchical representation from data. Overall, the LSTM + XGBoost hybrid model is the most promising model and deserves further research and potential development to improve its anti-advertising effectiveness. The evaluation metrics of models used in the chapter are shown in Table 9.4.

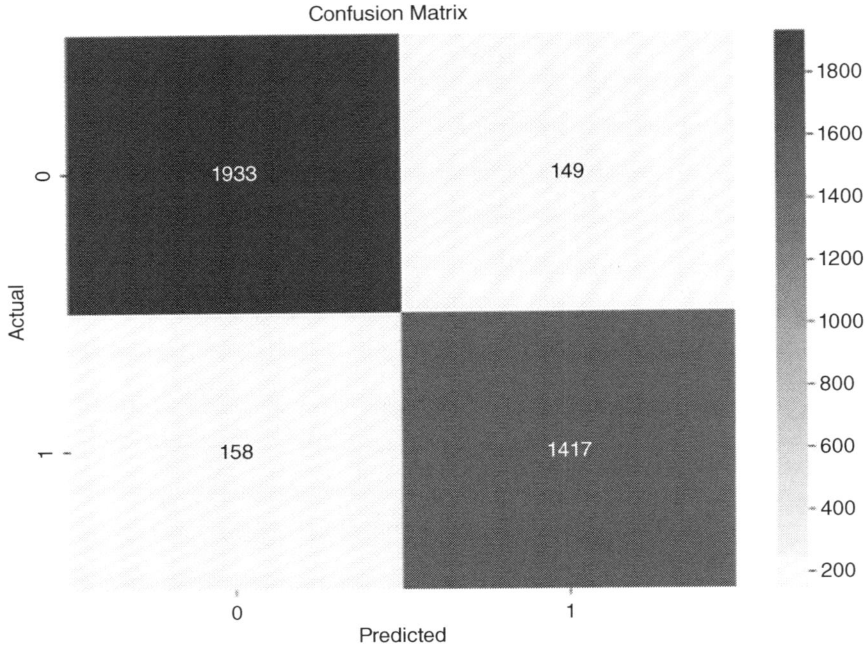

FIGURE 9.4 CNN model confusion matrix.

TABLE 9.4
Evaluation Metrics Comparison of Different Models

Modal	Precision	Recall	F1 Score	Accuracy
CNN	90.48	89.96	90.22	91.60
Random Forest	95.99	82.09	88.50	90.81
Naive Bayes	99.00	33.90	50.59	71.47
LSTM+XGBoost	93.74	92.14	92.93	93.96

The Confusion Matrix of Proposed model (LSTM+XGBoost) is shown in Figure 9.5. The significant Area Under the Curve (AUC) value of 0.98 on the Receiver Operating Characteristic (ROC) curve indicates that the XGBoost model performs exceptionally well in discriminating between positive and negative events in this Area of the slope of the curve, calculated as 0.98, indicates excellent prediction accuracy and longevity of the model. When the AUC is close to the optimal value of 1, this result indicates the most accurate classification model capable of providing accurate and reliable predictions, which highlights the improvement of the combined LSTM and XGBoost algorithms to emphasize your research.

The Receiver Operating Characteristics (ROC) Curve is shown in Figure 9.6.

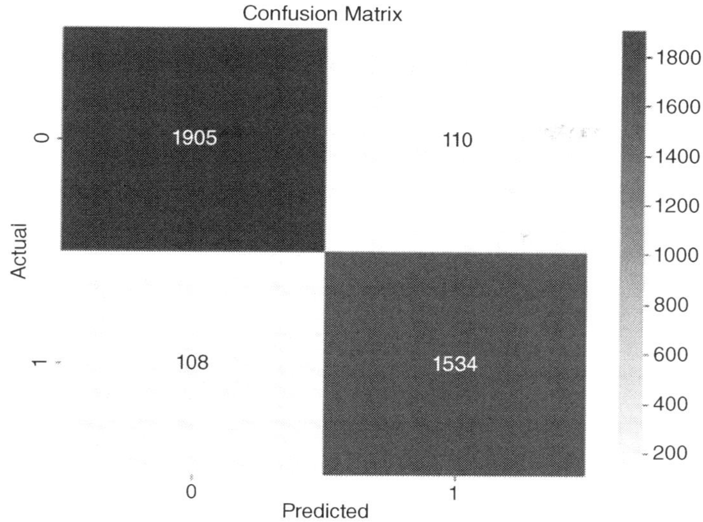

FIGURE 9.5 Confusion matrix of proposed model (LSTM+XGBoost).

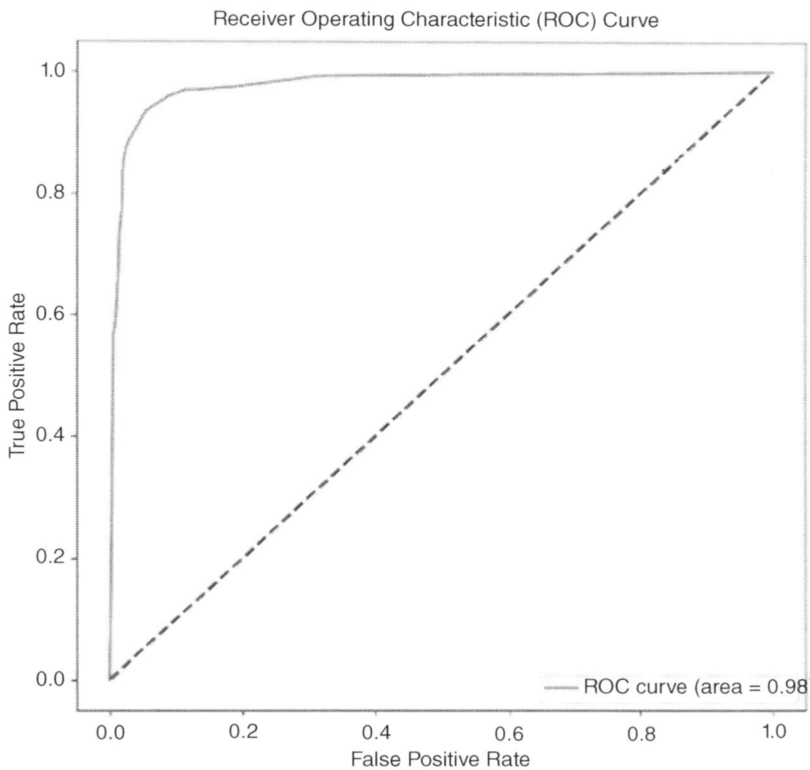

FIGURE 9.6 Receiver operating characteristics (ROC) curve.

9.8 CONCLUSION

This chapter investigates the effectiveness of LSTM and XGBoost working together in detecting fake news. The results show that the hybrid LSTM-XGBoost model achieves higher accuracy, precision, and recovery than the standard and baseline models. The main results show a 12–15% increase in accuracy compared to the same algorithm. The study shows how false positive classification can be achieved by combining XGBoost's ability to analyze random data with LSTM's ability to identify time trends. Patterns and features found by the LSTM layer help XGBoost distinguish true stories from false data. In conclusion, the LSTM-XGBoost combination provides a solid foundation for combating fake news. It has turned its attention to media integrity to prevent business interference, advertising, and online misrepresentation.

Going forward, research on fake news detection may discover more effective ways to supplement existing models. An important lesson involves the use of advanced language processing techniques (NLP) to capture complex words, thereby improving the model's ability to understand concepts and emotions. Moreover, combining deep learning methods such as Transformers, especially BERT and its variants, can achieve high performance by enabling both learning and mixed content. Additionally, the search for an integrated strategy to combine different models by trying to increase the power of each model will lead to stronger and more accurate predictions. Overcoming the challenge of changing the fake news model using adaptive learning and constantly updating the model with new information will help ensure the accuracy and effectiveness of the model over time. Finally, it is important to have a deep understanding of the AI identifier method to understand predictive models and support disclosure, especially in sensitive applications such as search, fake news, etc., so users trust and understand the decision-making process.

REFERENCES

Ahmed Sedik, Amr A. Abohany, Karam M. Sallam, Kumudu Munasinghe, T. Medhat, Deep fake news detection system based on concatenated and recurrent modalities, *Expert Systems with Applications*, 208, 2022, 117953. https://doi.org/10.1016/j.eswa.2022.117953

Alessandro Bondielli, Francesco Marcelloni, A survey on fake news and rumour detection techniques, *Information Sciences*, 497, 2019, 38–55. https://doi.org/10.1016/j.ins.2019.05.035

Anshika Choudhary, Anuja Arora, Assessment of bidirectional transformer encoder model and attention based bidirectional LSTM language models for fake news detection, *Journal of Retailing and Consumer Services*, 76, 2024, 103545, ISSN 0969-6989, https://doi.org/10.1016/j.jretconser.2023.103545

Argus Antonio Barbosa Cavalcante, Paulo Márcio Souza Freire, Ronaldo Ribeiro Goldschmidt, Claudia Marcela Justel, Early detection of fake news on virtual social networks: A time-aware approach based on crowd signals, *Expert Systems with Applications*, 247, 2024, 123350, ISSN 0957-4174, https://doi.org/10.1016/j.eswa.2024.123350

Choudhry A., I. Khatri, M. Jain, D. K. Vishwakarma, An emotion-aware multitask approach to fake news and rumor detection using transfer learning, *IEEE Transactions on Computational Social Systems*, 11(1), Feb. 2024, 588–599, https://doi.org/10.1109/TCSS.2022.3228312

Stephen C. Craig, Jason Gainous. 2024. To vote or not to vote? Fake news, voter fraud, and support for postponing the 2020 U.S. Presidential Election. *Politics & Policy*, 1–18, https://doi.org/10.1111/polp.12577

Daneshvar, B., Abdi, A., Hosseini, S.M. (2024). FNMD: An evaluation of machine learning and deep learning techniques for fake news detection. In: Santosh, K., et al. *Recent trends in image processing and pattern recognition. RTIP2R 2023. Communications in computer and information science*, vol. 2026, Springer, Cham, https://doi.org/10.1007/978-3-031-53082-1_22

Dev Deepali Goyal, Vishal Bhatnagar, Bhoopesh Singh Bhati, Manoj Gupta, Aziz Nanthaamornphong, LSTMCNN: A hybrid machine learning model to unmask fake news, *Heliyon*, 10, 3, 2024, e25244, ISSN 2405-8440, https://doi.org/10.1016/j.heliyon.2024.e25244

Eduri Raja, Badal Soni, Candy Lalrempuii, Samir Kumar Borgohain, An adaptive cyclical learning rate based hybrid model for Dravidian fake news detection, *Expert Systems with Applications*, 241, 2024, 122768, ISSN 0957-4174, https://doi.org/10.1016/j.eswa.2023.122768

Ercan Canhasi, Rexhep Shijaku, and Erblin Berisha. 2022. Albanian fake news detection, *ACM Trans. Asian Low-Resour. Lang. Inf. Process*, 21(5), September 2022, 86, 24 pages, https://doi.org/10.1145/3487288

Faramarz Farhangian, Rafael M.O. Cruz, George D.C. Cavalcanti, Fake news detection: Taxonomy and comparative study, *Information Fusion*, 103, 2024, 102140, ISSN 1566-2535, https://doi.org/10.1016/j.inffus.2023.102140

K. Ganesh Khekare Pavan Kumar, Kundeti Naga Prasanthi, Sanjiv Rao Godla, Venubabu Rachapudi, Mohammed Saleh Al Ansari and Yousef A. Baker El-Ebiary, Optimizing network security and performance through the integration of hybrid GAN-RNN models in SDN-based access control and traffic engineering, *International Journal of Advanced Computer Science and Applications(IJACSA)*, 14(12), 2023, http://doi.org/10.14569/IJACSA.2023.0141262

Z. Guo, K. Yu, A. Jolfaei, G. Li, F. Ding, A. Beheshti, Mixed graph neural network-based fake news detection for sustainable vehicular social networks, *IEEE Transactions on Intelligent Transportation Systems*, 24(12), Dec. 2023, 15486–15498, https://doi.org/10.1109/TITS.2022.3185013

Haosen Wang, Pan Tang, Hanyue Kong, Yilun Jin, Chunqi Wu, Linghong Zhou, DHCF: Dual disentangled-view hierarchical contrastive learning for fake news detection on social media, *Information Sciences*, 645, 2023, 119323, ISSN 0020-0255, https://doi.org/10.1016/j.ins.2023.119323

Izotova, O.A., Lavrova, D.S. Fake Post Detection Using Graph Neural Networks. *Automatic Control and Computer Sciences 55*, 1215–1221 (2021). https://doi.org/10.3103/S0146411621080393

Khekare, G., Midhunchakkravarthy, Smart Image Recognition System for The Visually Impaired People, *2023 International Conference on Energy, Materials and Communication Engineering (ICEMCE)*, Madurai, India, 2023, pp. 1–6, https://doi.org/10.1109/ICEMCE57940.2023.10434130

Khekare, Ganesh, Pushpneel Verma, Seema Raut, The smart accident predictor system using Internet of Things, *Cloud IoT*, 2022, Chapman and Hall/CRC, 163–175.

Khekare, G., Verma, P., Prophetic probe of accidents in Indian smart cities using machine learning. In V. Bhateja, S.C. Satapathy, C.M. Travieso-González, V.N.M. Aradhya (Eds.), *Data engineering and intelligent computing. Advances in intelligent systems and computing*, vol 1407, 2021, Springer, Singapore. https://doi.org/10.1007/978-981-16-0171-2_18

Khekare G., C. Masudi, Y. K. Chukka, D. P. Koyyada, Text normalization and summarization using advanced natural language processing, *2024 International Conference on Integrated Circuits and Communication Systems (ICICACS)*, Raichur, India, 2024, pp. 1–6, https://doi.org/10.1109/ICICACS60521.2024.10498983

G. K. Koru, Ç. Uluyol, Detection of Turkish fake news from Tweets with BERT models, *IEEE Access*, 12, 2024, 14918–14931, https://doi.org/10.1109/ACCESS.2024.3354165

K. Maity, S. Bhattacharya, S. Phosit, S. Kongsamlit, S. Saha, K. Pasupa, HANCaps: A two-channel deep learning framework for fake news detection in Thai. In B. Luo, L. Cheng, Z. G. Wu, H. Li, C. Li (eds) *Neural information processing. ICONIP 2023. Communications in computer and information science*, vol 1969, 2024, Springer, Singapore. https://doi.org/10.1007/978-981-99-8184-7_16

C. A. Maningo, S. P. A. M. Hlaing, Digitalization amidst educational crises: A comparative study of the new media literacy (NML) levels of the Philippines and Myanmar. *Asian Politics and Policy*, 16(1), 2024, 30–56. https://doi.org/10.1111/aspp.12728

S. I. Manzoor, J. Singla, Nikita, Fake news detection using machine learning approaches: A systematic review, *2019 3rd International Conference on Trends in Electronics and Informatics (ICOEI)*, Tirunelveli, India, 2019, pp. 230–234, https://doi.org/10.1109/ICOEI.2019.8862770

S. Mehravaran, P. Shamsinejadbabaki, Devising a machine learning-based instagram fake news detection system using content and context features, *Iran J Sci Technol Trans Electr Eng*, 47, 2023, 1657–1666, https://doi.org/10.1007/s40998-023-00635-4

Monther Aldwairi, Ali Alwahedi, Detecting fake news in social media networks, *Procedia Computer Science*, 141, 2018, 215–222, ISSN 1877-0509, https://doi.org/10.1016/j.procs.2018.10.171

Qian Chang, Xia Li, Zhao Duan, Graph global attention network with memory: A deep learning approach for fake news detection, *Neural Networks*, 172, 2024, 106115, ISSN 0893-6080, https://doi.org/10.1016/j.neunet.2024.106115

Rohit Kumar Kaliyar, Anurag Goswami, Pratik Narang, Soumendu Sinha, FNDNet – A deep convolutional neural network for fake news detection, *Cognitive Systems Research*, 61, 2020, 32–44, ISSN 1389-0417, https://doi.org/10.1016/j.cogsys.2019.12.005

P. Santhiya, S. Kavitha, T. Aravindh, S. Archana, A. V. Praveen, Fake news detection using machine learning, *2023 International Conference on Computer Communication and Informatics (ICCCI)*, Coimbatore, India, 2023, 1–8, https://doi.org/10.1109/ICCCI56745.2023.10128339

Xiaochang Fang, Hongchen Wu, Jing Jing, Yihong Meng, Bing Yu, Hongzhu Yu, Huaxiang Zhang, NSEP: Early fake news detection via news semantic environment perception, *Information Processing & Management*, 61(2), 2024, 103594, ISSN 0306-4573, https://doi.org/10.1016/j.ipm.2023.103594

Xinyi Zhou, Atishay Jain, Vir V. Phoha, Reza Zafarani. Fake news early detection: A theory-driven model. *Digital Threats*, 1(2), (June 2020), 12, 25 pages, https://doi.org/10.1145/3377478

Y. Yu, X. Si, C. Hu, J. Zhang, A review of recurrent neural networks: LSTM cells and network architectures, *Neural Computation*, 31(7), July 2019, 1235–1270, https://doi.org/10.1162/neco_a_01199

X. Zhang, S. Dadkhah, A. G. Weismann, M. A. Kanaani, A. A. Ghorbani, Multimodal fake news analysis based on image–text similarity, *IEEE Transactions on Computational Social Systems*, 11(1), Feb. 2024, 959–972, https://doi.org/10.1109/TCSS.2023.3244068

10 An In-Depth Comparative Analysis of Malware Classification Techniques

Dipesh Vaya and Rakesh Kumar Saxena

10.1 INTRODUCTION

The domain of cybersecurity is developing with its increasing needs for society and the significance of malware detection and classification stands as a pivotal challenge. With the rise in use and technology, malware threats are also evolving rapidly. To counter this, it is necessary to mitigate the gaps between research and practical solutions and continue improvements in the techniques of classifying malware threats. Hence the need for more sophisticated, reliable, robust, adaptable, and scalable classification techniques is a must. Therefore, this research study aims to churn out the substantial study of malware classification, offering a comprehensive review and comparative analysis of existing models through a novel perspective.

The inception of malware classification models marked a significant milestone in cybersecurity. These models were designed to identify and categorize malicious software, thereby enabling better defense mechanisms against cyber threats. However, as the complexity of malware has escalated, so too has the challenge of effectively classifying it. Traditional models, while foundational, often struggle to keep pace with the rapid advancements and diversifications in malware (Shashi et al., 2023; Vaya & Hadpawat, 2021). In the era of IoE, networks, processes, people, and the internet (Vaya & Hadpawat, 2020) cybersecurity plays a vital role in protecting data. This inadequacy manifests in precision, accuracy, scalability, and adaptability limitations. Furthermore, the field has witnessed a proliferation of models, each claiming superiority in certain aspects but often lacking a comprehensive evaluation against a diverse set of metrics.

Literature reviews presented by many researchers till now focused only on some adjectives like speed or accuracy of malware threat classification methods; therefore, it falls inadequate in understanding the performance of such methods. Limitations and other constraints like recall, complexity, cost, and delay also play an important role in comparing the performance of threat classification methods. This review paper is focused on both aspects of comparison; one is performance, and the other is tradeoffs to provide a balanced and holistic assessment of methods for classifying

DOI: 10.1201/9781003498094-10

malware threats. This paper also introduces a novel ranking of the methods reviewed for providing a unified framework for evaluating and comparing malware classification models. This paper will work as a torch for further researchers to enhance cybersecurity defenses against the dynamic and complex landscape of malware threats.

10.2 REVIEW OF LITERATURE

Researchers have proposed a wide variety of models for classifying malware threats in different network scenarios. The review of literature provided in this section showcases a detailed overview of various research studies related to malware classification, upholding a wide range of topics within the field of cybersecurity. These research studies define the ever-evolving security challenges posed by malicious software and aim to develop innovative approaches to counteract them. The study (Gulatas et al., 2023) brings out the point that securing the Internet of Things (IoT) devices in the edge/fog computing environment is very important since the security of these devices and scenarios is uncertain. A study that identifies and characterizes 64 families of zero-day malware operating in IoT target architectures is carried out, informing about delivery methods and attack vectors to help guide future research in the domain of IoT security.

A framework is proposed in the study (Zhong et al., 2023) for the visualized classification of malware by means of converting them into images and applying image enhancement techniques; this is called VisMal. The framework shows promising accuracy and speed, thus presenting security engineers with a practical tool in the domain of malware classification. The work (Djafer Yahia et al., 2023) presents a subspace-based method with respect to the classification and visualization of malware files through a subspace representation of malware image patterns. The proposed approach significantly outperforms the previous methods, and hence, the promising solution for detecting advanced malware threats is displayed. The work (Zhang et al., 2023) introduces a lightweight and low-cost machine learning-based malware traffic classification method to be applied on IoT devices and IoT-related scenarios. The broad learning (BL) used in the proposed method exceeds results but also introduces lower computational overhead, which therefore makes it a good alternative to IoT security. Challenging research concerning recognizing Android malware families in a deep open-world setting is discussed in the paper (Lu & Wang, 2024). The proposed solution, DOMR, succeeds in obtaining an accurate classification in deep open-world scenarios. This makes the model very effective while implementing the malware on a wide variety of malware families. Another research (Kim et al., 2023) introduces a CNN-based classification method to malware through non-disassembled binary files that associates two modalities: "malware images" and "structural entropies." The model can yield higher accuracy while keeping the disassembly out of the process, and hence, it is a good alternative for the process of malware classification. In Li et al. (2023), a multimodal fusion and weight self-learning approach for malware family classification is proposed, effectively fusing different modalities of malware data samples. In this context, the approach has exhibited excellent classification performance, thus contributing toward the identification and classification of malware families in Intelligent Transportation Systems.

With the increasing number of Android malware security tools that allow swift identification and categorization of various malware classes to formulate rapid response strategies (Almarshad et al., 2023), there is a need to augment these security solutions with machine learning and deep learning methods. However, one significant drawback has been the lack of data needed for training the deep learning models. In this regard, the proposed one-shot learning-based Siamese neural network is designed to effectively detect and categorize Android malware attacks. In the dataset of Drebin, which consists of both benign and harmful components, the approach uses the proposed method. A demonstration of five major phases of the architecture employed, model architecture, data partitioning, training, and evaluation was found to produce a high of 98.9% accuracy, superior to all the other standard methods.

As 5G technology has achieved prominence in the Industrial Internet of Things (IIoT), it integrates artificial intelligence and cloud computing for low-latency integration, enhancing industrial processes (Ahmed et al., 2023). However, with the increasing advancements in 5G technology, they also bring about new security risks and attack vectors in the domain of malware attacks on highly connected IoT devices and scenarios. A deep learning-based architecture for classifying malware attacks on IIoT is designed in the approach, using image representations of malware and convolutional neural networks (CNN) to differentiate various malware attacks, attaining an accuracy level of 97%.

Malware traffic classification has been witnessed to advance into deep learning techniques, especially with the increasing volume of network traffic data (Barut et al., 2023). To mitigate privacy concerns through the usage of payload data and thus improve model generalization, an introduction of a Residual 1-D Image Transformer (R1DIT) system is proposed. This system thus uses domain knowledge by parsing network traffic headers and raw data transformation and attention-based modules to classify malware types and benign traffic. R1DIT achieved better performance than other methods with a 0.972 F1 score on the NetML dataset and a 0.9999 F1 score for multi-class malware classification using the CICIDS2017 dataset, 20 samples. Wong et al. (2024) introduced APILI: a behavior-based malware analysis method using deep learning. This method utilizes a deep learning algorithm to locate API calls related to malware techniques. It also incorporates MITRE ATT&CK framework for enhancing behavior-based malware detection and analysis by locating low-level API calls corresponding to high-level malicious behaviors.

Malware recognition faces challenges, both in finding known and unknown malware families in real-world scenarios. Thus, the need for malware open-set recognition (MOSR) (Guo et al., 2023) is that existing approaches make malware families known to the classifier, while in fact, there are no known malware families in the test set. In view of these findings, a novel model is thereby suggested that synthesizes malware instances mimicking unknown malware families. The paper describes a model built from generative adversarial networks, making robust training to improve MOSR performance levels.

In Chen et al. (2023), the major concern of wireless networks is also undertaken by introducing a hypergraph-based model that describes the propagation of malware across large-scale wireless networks. The paper explains several parameters like the number of devices in the network and the type of malware as a criterion of determination for the malware outbreak threshold. The study, however, reveals that isolation connections on the internet do not entirely curtail malware outbreaks, especially in scenarios where these networks feature wireless connections. In Beg et al. (2023), the research proposes to develop an augmented convolutional model (ACM) with forensic neural networks (FNNs) for cross-domain malware analysis. The suggested ACM showed high accuracy in categorizing various malware datasets and demonstrated its real-time potential for malware analysis. The ACM showed promise in localizing malware locations within data streams. He et al. (2023) introduced a high-speed, scalable phylogenetic tree construction algorithm for malware clustering. The proposed approach will decrease the construction cost of phylogenetic trees, hence improving clustering accuracy. The study also describes an online processing algorithm to add new malware specimens to existing phylogenetic trees sequentially to be more efficient when processing large-scale malware sets. An in-depth summary of these methods can be observed in Table 10.1.

Khan et al. (2023) introduced a few-shot learning approach for Android malware detection using a Relation Network architecture. Proposed architecture involves the use of discriminative feature embedding and relation modules to improve accuracy even when the number of samples is limited. Dou et al. (2024) focused on the modeling of malware propagation between wearable IoT devices and scenarios. It uses cellular automata and models of human mobility to capture the dynamics of malware spread. The study provides insights into the impact of various parameters on botnet formation in wearable IoT networks.

Integrating malware classification techniques into knowledge management systems (KMS) presents several challenges as identified in the literature. Signature-based techniques struggle with adaptability, requiring frequent updates to maintain relevance, while behavior-based approaches face issues in converting complex, dynamic data into structured knowledge. Heuristic and rule-based methods necessitate constant rule updates, which can overwhelm the KMS and compromise consistency. Machine learning-based classification introduces the challenge of managing high-dimensional data and model retraining, placing additional demands on KMS infrastructure. Hybrid techniques, which combine multiple approaches, further complicate integration by generating diverse data types and requiring synchronized updates, posing difficulties in maintaining a cohesive and scalable knowledge management framework. These challenges highlight the complexities of merging cybersecurity tools with knowledge management practices, necessitating adaptations in both fields to ensure effective integration (Ogala et al., 2023).

10.3 RESULTS

This section discusses a comprehensive overview of various malware classification methods, each evaluated across six key criteria: Precision, Accuracy, Recall, Delay, Cost, and Scalability. The proposed ranking metric reflects a holistic, balanced approach toward the evaluation of malware classification methods. Unlike the traditional methods, which are usually confined to an independent look at performance metrics like precision, accuracy, or recall, this metric addresses the multitasking nature of real-world applications through incorporating several key performance

TABLE 10.1
Empirical Analysis of the Reviewed Methods

Ref. No.	Method Name	Findings	Advantages	Limitations	Impacts
1	Deep Learning for 5G-enabled IIoT Malware Classification	Designs a deep learning-based architecture to classify malware attacks on 5G-enabled IIoT networks, achieving 97% accuracy. Utilizes image representation and convolutional neural networks.	Enables effective classification of malware attacks on 5G-enabled IIoT networks. Achieves a high accuracy rate of 97%. Enhances security in IIoT ecosystems.	Limited discussion on the scalability of the proposed system. No exploration of false positives.	Enhances security in 5G-enabled IIoT networks, ensuring the integrity of industrial procedures and data privacy.
2	Siamese Shot for Android Malware Detection	Using the Drebin dataset, a Siamese neural network-based approach is proposed for Android malware detection and categorization. Achieved an accuracy of 98.9%, outperforming other standard methods.	Overcomes data scarcity in malware samples. Efficiently identifies and categorizes malware. Uses Siamese networks for similarity ranking. Provides high accuracy (98.9%).	Scarcity of data for malware samples. Limited to the Drebin dataset. No discussion of real-world deployment.	Enhances Android malware detection with high accuracy, potentially improving device security.
3	R1DIT: Malware Traffic Classification using Raw Data	Introduces R1DIT model for malware traffic classification. Leverages network domain knowledge and raw data samples. Achieves high F1 scores on NetML and CICIDS2017 datasets. Generalizes well to new traffic types.	Utilizes raw traffic data for malware classification. Addresses privacy concerns and generalization issues. Achieves high F1 scores compared to feature-based methods. Performs well with TLS 1.3 traffic.	Does not discuss the computational resources required. Lacks analysis of potential false positives.	Enhances malware traffic classification using raw data, addressing privacy concerns and improving model generalization.
4	Augmented Convolutional Model for Malware Analysis	Introduces an augmented convolutional model (ACM) for cross-domain malware analysis using forensic neural networks (FNNs). Achieves high accuracy in classifying malware and localizing it in data streams. Outperforms recent models in both aspects.	Proposes an ACM for intelligent cross-domain malware analysis. Achieves high accuracy in malware classification and localization. Outperforms recent models in multiple malware datasets.	Does not discuss real-time model deployment or scalability. Limited to malware detection and localization.	Provides a high-accuracy solution for cross-domain malware analysis, aiding in real-time detection and localization for improved cybersecurity.

(Continued)

TABLE 10.1
(Continued)

Ref. No.	Method Name	Findings	Advantages	Limitations	Impacts
5	Hypergraph-Based Model for Wireless Network Malware	Proposes a hypergraph-based model for describing malware propagation over large-scale wireless networks. Identifies malware outbreak threshold. Shows that malware pandemic depends on network size and malware characteristics. Suggests that isolating Internet connections doesn't fully prevent malware spread. Demonstrates real data validation.	Provides a novel approach for modeling malware propagation over wireless networks. Offers insights into network size and malware's nature's impact on outbreaks. Validates findings with real data samples.	Does not discuss practical implementation or resource requirements. Focused on theoretical analysis.	Provides a theoretical foundation for understanding malware propagation in large-scale wireless networks, aiding in network security strategies.
6	Subspace-Based Malware Classification	Proposal of a subspace-based approach for classifying malware files using subspace representation and visualization.	Detailed and precise approach, effective detection of emerging malware.	Performance evaluation on specific datasets and potential vulnerabilities.	Improved detection of advanced malware.
7	Modeling IoT Malware Propagation with Cellular Automata	Proposes an individual-based cellular automata (CA) model for studying malware propagation in wearable IoT devices & scenarios. Utilizes a human mobility model for realistic scenarios. Achieves precise modeling of malware propagation.	Introduces an individual-based CA model for wearable IoT malware propagation. Utilizes a human mobility model for realism. Achieves precise malware propagation modeling.	Does not discuss practical deployment of the CA model. Limited to IoT malware propagation in wearable devices & scenarios.	Provides a precise model for malware propagation in wearable IoT devices, aiding in developing targeted defense strategies.
8	MOSR: Open-set Malware Recognition	Addresses the challenge of open-set malware recognition. Proposes a model that synthesizes malware instances to mimic unknown malware families. Achieves robust recognition of unknown malware.	Tackles open-set malware recognition, including unknown families. Synthesizes malware instances for better recognition. Achieves robust recognition of unknown malware.	Limited discussion of real-world deployment scenarios. Lacks exploration of false positives and false negatives.	Enhances malware recognition by addressing the challenge of unknown malware families, potentially improving cybersecurity.

9	IoT Malware Analysis	Characterizing 64 IoT malware families, understanding their behaviors, and adopting investigation frameworks.	In-depth analysis of IoT malware, identification of attack vectors, informative for future research.	Limited access to malware samples, and constrained defense capabilities.	Enhanced understanding of IoT malware threats.
10	Scalable Phylogenetic Tree Construction for malware clustering	Proposes a high-speed, scalable phylogenetic tree construction algorithm with a clustering algorithm for IoT malware. Reduces computation by 97.52% compared to conventional methods. Achieves high clustering accuracy.	Introduces a fast and scalable phylogenetic tree construction method for IoT malware. Reduces computational cost significantly. Achieves high clustering accuracy for different scenarios.	Does not discuss practical application or deployment scenarios. Limited to IoT malware clustering.	Enhances IoT malware clustering by significantly reducing computational cost and improving clustering accuracy, aiding in cyber defense.
11	Few-Shot Learning-Based Malware Detection	Proposes a Relation Network-based architecture for few-shot malware detection. Achieves high accuracy even with limited training samples. Outperforms baseline methods significantly.	Introduces a novel architecture based on Relation Network for few-shot malware detection. Achieves high accuracy with limited training samples. Outperforms baseline methods by a large margin.	Does not discuss the challenges of applying few-shot learning in practical scenarios. Focused on malware detection using few-shot learning.	Enhances malware detection with few-shot learning, providing improved accuracy with limited training samples and aiding in rapid threat detection.
12	CNN-Based Malware Classification	Proposal of a CNN-based malware classification model using non-disassembled binary files.	Improved accuracy, no disassembly required, effective association of modalities.	Dependency on binary files, potential limitations in disassembly, specificity to certain datasets.	Efficient and accurate malware family classification.
13	Multimodal Malware Classification	Introduction of a malware family classification approach based on multimodal fusion and weight self-learning.	Excellent classification performance, efficiency, and small resource overhead.	Imbalanced datasets, and limited technical details.	Enhanced malware family identification.

(Continued)

TABLE 10.1
(Continued)

Ref. No.	Method Name	Findings	Advantages	Limitations	Impacts
14	Deep Open-World Android Malware Recognition	Introduction of a deep learning-based approach (DOMR) for open-world Android malware family recognition.	Addresses open-world challenges, incremental update capability, and outperforms existing methods.	Closed-world assumption, specialized datasets, potential limitations in practical applications.	Improved Android malware family recognition.
18	APILI: Behavior-based Malware Analysis using Deep Learning	Introduces APILI, a behavior-based malware analysis approach using deep learning. Locates API calls corresponding to malware techniques. Outperforms traditional and ML-based techniques in technique discovery and API locating.	Utilizes deep learning to locate API calls related to malware techniques. Incorporates MITRE ATT&CK framework. Outperforms traditional methods in technique discovery and API locating.	No discussion of model deployment in practical settings. Limited to malware behavior analysis.	Enhances behavior-based malware analysis by locating low-level API calls corresponding to high-level malicious behaviors, improving malware detection and analysis.
19	Broad Learning for Malware Traffic Classification	Proposal of a lightweight and cost-effective malware traffic classification method (BL-MTC) using broad learning.	Superior results, exponential advantage in computation overheads.	Limited to the USTC-TFC2016 dataset, limited technical details.	Effective and low-cost malware traffic classification.
20	VisMal	Introduction of a visualized malware classification framework (VisMal) for efficient categorization using image representations.	Fast classification, acceptable accuracy, simple visualization approach.	Limited evaluation, reliance on image-based representation.	Efficient and accurate malware classification.

indicators into one single framework. To be precise, the ranking metric combines precision, accuracy, recall, delay, cost, and scalability, all important critical dimensions in the effectiveness and practicality assessment of malware classification models. This means that precision ensures the minimization of false positives, while accuracy measures overall correctness, and recall evaluates the model for the identification of actual malware. While delay is very crucial for real-time applications where speed detection is highly critical, cost depicts computational and resource overheads that may affect the applicability of a model in large-scale or resource-constrained environments. Whereas scalability refers to how well performance scales when results are extrapolated onto greater data or across other various platforms—a crucial, growing aspect of malware datasets in cloud or IoT environments. Altogether, this allows the ranking framework to make more holistic evaluations. That is, it represents not just the isolated strengths of each model but its overall suitability to diverse cybersecurity applications. These metrics are depicted in Table 10.2 and are crucial for assessing the performance and practicality of these methods in real-world scenarios.

TABLE 10.2
Comparative Analysis of the Reviewed Methods

Ref. No.	Method Name	Precision	Accuracy	Recall	Delay	Cost	Scalability
1	Deep Learning for 5G-enabled IIoT Malware Detection	F	M	M	G	G	F
2	Siamese Shot for Android Malware Detection	E	G	M	E	G	E
3	R1DIT: Malware Traffic Classification using Raw Data	E	G	F	G	M	G
4	Augmented Convolutional Model for Malware Analysis	G	P	M	G	F	F
5	Hypergraph-Based Model for Wireless Network Malware	F	P	E	E	E	G
6	Subspace-Based Malware Classification	P	F	G	E	E	E
7	Modeling IoT Malware Propagation with Cellular Automata	M	M	G	E	G	G

(Continued)

TABLE 10.2
(Continued)

Ref. No.	Method Name	Precision	Accuracy	Recall	Delay	Cost	Scalability
8	MOSR: Open-set Malware Recognition	G	G	F	G	E	F
9	IoT Malware Analysis	P	F	G	M	P	P
10	Scalable Phylogenetic Tree Construction for Malware	M	M	E	P	M	M
11	Few-Shot Learning-Based Malware Detection	G	P	P	F	P	P
12	CNN-Based Malware Classification	E	M	M	P	M	E
13	Multimodal Malware Classification	F	G	M	G	M	P
14	Deep Open-World Android Malware Recognition	E	F	E	G	M	E
18	APILI: Behavior-based Malware Analysis using Deep Learning	G	P	M	F	P	E
19	Broad Learning for Malware Traffic Classification	F	G	P	F	M	F
20	VisMal	P	P	G	M	M	F

Notes: E stands for Excellent (Very High), G stands for Good (High), M stands for Medium, F stands for Fair (Low), and P stands for Poor (Very Low)

The selection of method can be done as per the application by using a weighted scoring approach in which the importance of each metric can be adjusted regarding a particular use case or application requirement.

Selection Score = $w_1 \cdot P_i + w_2 \cdot A_i + w_3 \cdot R_i + w_4 \cdot D_i + w_5 \cdot C_i + w_6 \cdot S_i$

Where:

P_i is the precision score of the model, A_i is the accuracy score, R_i is the recall score, D_i is the delay score, C_i is the cost score, S_i is the scalability score, and w_1, w_2 …, w_6 are the weights assigned to each criterion based on the application's priority.

For example, in IoT malware detection, where the need for speed and scalability of huge volumes is very important, latency, and scalability could be preferred over precision; in high-security environments, however, where a false positive can cost a lot, such as financial systems, precision may assume more weight. This flexibility allows the ranking metric to be adjusted to a wide range of situations and gives, in general, an extremely useful tool in the selection and evaluation of models in malware classification when operating real-world tasks. Aggregated scores are computed, in most cases, by using a linear weighted sum, but other nonlinear methods can be used when some metrics are more critical than others. The ranking brings out a rich picture of those models that result in good performance both in isolated metrics and in their harmonized combinations for performance criteria. This holistic malware classification evaluation carries important implications for future research in the development of more balanced models to meet complex and changing cybersecurity challenges. It is a ranking metric that balances different key performance indicators and guides the model selection not only for ideal condition performance but also toward practical usefulness in realistic settings.

As per Table 10.2, a detailed analysis of the methods, highlighting their strengths and weaknesses in each criterion, can be discussed as follows:

- **Deep Learning for 5G-enabled IIoT Malware Detection:** High Accuracy and Recall but with a high Delay. Suitable for industrial IoT security where precision is crucial.
- **Siamese Shot for Android Malware Detection:** Achieves a balance between Precision and Recall with high Scalability. It's suitable for various applications.
- **R1DIT: Malware Traffic Classification using Raw Data:** High Accuracy and Scalability, but with a high Delay. Suitable for network-level analysis.
- **Augmented Convolutional Model for Malware Analysis:** High Recall but low Precision, limiting its accuracy.
- **Hypergraph-Based Model for Wireless Network Malware:** High Recall and Scalability but with low Accuracy. Suitable for modeling malware propagation.
- **Subspace-Based Malware Classification:** This method offers high Recall and Scalability but at the cost of low Precision and Accuracy. It may be suitable for large-scale deployments where false positives can be managed.
- **Modeling IoT Malware Propagation with Cellular Automata:** High Recall and Scalability, suitable for modeling IoT malware propagation.
- **MOSR: Open-set Malware Recognition:** High Precision and Scalability but with a high Delay. Suitable for open-set malware recognition.
- **IoT Malware Analysis:** Although it shows high Recall, it has low Precision and Scalability. This method is valuable for comprehensive analysis but may result in more false positives.
- **Scalable Phylogenetic Tree Construction for Malware:** High Recall and Scalability, making it suitable for malware clustering.
- **Few-Shot Learning-Based Malware Detection:** Balanced Precision and Recall with Low Delay. Suitable for few-shot learning applications.
- **CNN-Based Malware Classification:** Offers a good balance between Precision and Recall with high Scalability. It may be suitable for a wide range of applications.

- **Multimodal Malware Classification:** High Recall and Precision make it effective but with high Delay. It's suitable for in-depth analysis but may not be practical for real-time detection.
- **Deep Open-World Android Malware Recognition:** High Recall and Scalability indicate its suitability for large-scale Android malware detection. However, it has a high Delay and Cost.
- **APILI: Behavior-based Malware Analysis using Deep Learning:** High Precision and Recall with Scalability. Effective for behavior-based malware analysis.
- **Broad Learning for Malware Traffic Classification:** High Accuracy and Scalability make it suitable for network-level analysis. However, it has a high Delay, limiting its real-time applicability.
- **VisMal:** With high Recall, it excels in comprehensive detection but may generate more false positives. Its low Delay and Cost make it practical for real-time analysis.

Several real-world examples illustrate the successful integration of malware classification techniques into knowledge management systems (KMS). For instance, IBM's QRadar platform integrates signature-based and machine learning techniques, combining real-time threat intelligence with knowledge systems for automated responses in organizations like healthcare and finance (Samtani et al., 2017). Similarly, CISCO's Talos Intelligence Group employs hybrid classification methods (signature, behavior, heuristic) to update threat intelligence across CISCO's products, widely used in sectors such as telecommunications (Cisco Talos Intelligence Group, 2019). Microsoft Defender ATP leverages machine learning and behavior-based techniques to feed security data into an evolving KMS, protecting large enterprises and governments (Egele et al., 2012). Symantec's Integrated Cyber Defense (ICD) platform uses behavior-based heuristics, and machine learning approaches to centralize threat intelligence across industries like retail and healthcare (Symantec by Broadcom, n.d.). Finally, CrowdStrike Falcon integrates machine learning and behavior-based classification into its platform to provide real-time security insights for a variety of businesses, particularly in finance (CrowdStrike, 2020. These cases highlight how organizations effectively merge cybersecurity tools with KMS to enhance their ability to detect and respond to evolving threats.

10.4 CONCLUSION

To conclude the importance and evaluation of the available methods of malware classification, metrics such as precision, accuracy, and recall are studied to assess the performance and efficiency. However, it is also important to focus on the tradeoffs in these methods, such as delay, cost, and scalability. By addressing the said factors, the research community can support informed decision-making in selecting and implementing classification methodologies and advancing the field with a comprehensive understanding of their strengths and limitations. Overall, the choice of a malware detection method depends on the specific use case and priorities. Methods like "Deep Open-World Android Malware Recognition" and "CNN-Based Malware

Classification" offer a balanced performance across multiple criteria and can be considered suitable for a wide range of applications. However, for specific needs like precision-focused detection or scalability, other methods may be more appropriate for real-time scenarios.

The precision, accuracy, and recall metrics offer insights into the ability of these methods to classify malware while minimizing false positives and false negatives accurately. It is evident that some methods, such as "Deep Open-World Android Malware Recognition," demonstrate very high precision and very high recall, and "CNN-Based Malware Classification" demonstrates very high precision, balanced accuracy, and recall, indicating their proficiency in malware detection. In the analysis, factors like delay, cost, and scalability are also considered, and it is found that methods like "CNN-Based Malware Classification" offer very low delay and very high scalability with moderate cost, making it efficient choices for real-time applications. Factors like scalability also play a vital role in malware classification, "R1DIT: Malware Traffic Classification using Raw Data" offers high accuracy and scalability but with a high delay. Similarly, some other methods studied in this study also offer high scalability with accuracy, recall, and precision tradeoffs. This study advances theoretical knowledge in the domain of malware classification. Existing models often focus on a single performance aspect, while this work proposes a holistic framework that evaluates the tradeoffs among various metrics. Additionally, the research enhances theoretical discussions by addressing the adaptability of malware classification models in dynamic environments, such as IoT and 5G networks, challenging the static assumptions found in traditional models. The study also contributes to the ongoing development of machine learning applications in cybersecurity, providing a foundation for future theoretical exploration of hybrid and AI-based models.

From a practical standpoint, this research provides valuable insights for cybersecurity professionals and organizations in selecting appropriate malware classification models for specific environments. These findings enable better decision-making regarding the deployment of classification models in complex settings. Organizations can use the findings to enhance their threat detection and mitigation strategies, ensuring robust security measures that can scale with growing data demands. Additionally, the adaptability of the models presented in this study makes them suitable for emerging technologies, including smart cities and autonomous systems, where traditional models may struggle to keep pace with evolving threats. This research offers scalable and practical solutions for industries facing increased cybersecurity challenges in the face of rapid technological advancements. It is important to understand the need and applicability of a particular method and should be aligned with the specific needs, limitations, and objectives of the domain. There cannot be a single solution or method that fits all situations. Therefore, it is imperative to carefully study methods and their applicability by comparing the tradeoffs between precision, efficiency, and scalability.

The novel ranking metric introduced in this research offers a practical tool for security professionals and organizations, allowing them to make informed decisions when selecting and deploying malware classification models. The comparative analysis of models based on metrics like precision, accuracy, scalability, and delay ensures that this research aids in developing more efficient and robust cybersecurity measures. Furthermore, by addressing the limitations of existing models, this research directly impacts cybersecurity strategies by enhancing malware detection and response capabilities in dynamic environments like IoT and 5G networks. This analytical research study will help the research community by offering a guide tailored to different domains and priorities for classifying malware attacks. This research study will guide cybersecurity experts in making informed decisions tailored to their specific needs.

The future scope of this study may include the continuous evolution of malware threats and ever-evolving technologies and tactics to detect new types of threats. This research study can be extended further by incorporating new innovative methods for detecting malware in different scenarios and cross domains, for example, IoT, IoE, 5G-enabled systems, fog computing, edge computing, etc., where tailored malware classification solutions are in high demand. Further study can explore the innovative fusion of machine learning and artificial intelligence in detecting malware threats; these methods can be compared based on performance metrics such as precision, accuracy, and recall along with tradeoffs in terms of scalability, speed of detection, resource optimization, and integration cost with other systems.

REFERENCES

Ahmed, M., Anisetti, M., Ahmad, A., & Jeon, G. (2023). A multilayer deep learning approach for malware classification in 5G-enabled IIoT. *IEEE Transactions on Industrial Informatics*, 19(2), 1495–1503. https://doi.org/10.1109/TII.2022.3205366

Almarshad, F. A., Zakariah, M., Gashgari, G. A., Aldakheel, E. A., & Alzahrani, A. I. A. (2023). Detection of Android malware using machine learning and Siamese shot learning technique for security. *IEEE Access*, 11, 127697–127714. https://doi.org/10.1109/ACCESS.2023.3331739

Barut, O., Luo, Y., Li, P., & Zhang, T. (2023). R1DIT: Privacy-preserving malware traffic classification with attention-based neural networks. *IEEE Transactions on Network and Service Management*, 20(2), 2071–2085. https://doi.org/10.1109/TNSM.2022.3211254

Beg, R., Pateriya, R. K., & Tomar, D. S. (2023). ACMFNN: A novel design of an augmented convolutional model for intelligent cross-domain malware localization via forensic neural networks. *IEEE Access*, 11, 87945–87957. https://doi.org/10.1109/ACCESS.2023.3305274

Chen, J., Sun, S., Xia, C., Shi, D., & Chen, G. (2023). *Modeling and analyzing malware propagation over wireless networks based on hypergraphs*. IEEE Transactions on Network Science and Engineering, 10(6), 3767–3778. https://doi.org/10.1109/TNSE.2023.3273184

Cisco Talos Intelligence Group (2019). *The impact of hybrid threat intelligence: A white paper*. Cisco. https://www.cisco.com/c/dam/global/en_sg/solutions/industries/talos_white_paper.pdf

CrowdStrike. (2020). Falcon complete: Data sheet. CrowdStrike. https://www.crowdstrike.com/wp-content/uploads/2020/02/crowdstrike-falcon-complete-data-sheet.pdf

Djafer Yahia, B. M., Batalo, B., & Fukui, K. (2023). Efficient malware analysis using subspace-based methods on representative image patterns. *IEEE Access*, 11, 102492–102507. https://doi.org/10.1109/ACCESS.2023.3313409

Dou, J., Xie, G., Tian, Z., Cui, L., & Yu, S. (2024). Modeling and analyzing the spatial-temporal propagation of malware in mobile wearable IoT networks. *IEEE Internet of Things Journal*, 11(2), 2438–2452. https://doi.org/10.1109/JIOT.2023.3295016

Egele, M., Scholte, T., Kirda, E., Kruegel, C., (2012). A survey on automated dynamic malware-analysis techniques and tools. *ACM Computing Surveys (CSUR)*, 44(2), 1–42. https://doi.org/10.1145/2089125.2089126

Guo, J., Guo, S., Ma, S., Sun, Y., & Xu, Y. (2023). Conservative novelty synthesizing network for malware recognition in an open-set scenario. *IEEE Transactions on Neural Networks and Learning Systems*, 34(2), 662–676. https://doi.org/10.1109/TNNLS.2021.3099122

Gulatas, H. H., Kilinc, A., Zaim, A. H., & Aydin, M. A. (2023). Malware threat on edge/fog computing environments from internet of things devices perspective. *IEEE Access*, 11, 33584–33606. https://doi.org/10.1109/ACCESS.2023.3262614

He, T., Han, C., Isawa, R., Takahashi, T., Kijima, S., & Takeuchi, J. (2023). Scalable and fast algorithm for constructing phylogenetic trees with application to IoT malware clustering. *IEEE Access*, 11, 8240–8253. https://doi.org/10.1109/ACCESS.2023.3238711

Khan, F. B., Durad, M. H., Khan, A., Khan, F. A., Chauhdary, S. H., & Alqarni, M. (2023). Detection of data scarce malware using one-shot learning with relation network. *IEEE Access*, 11, 74438–74457. https://doi.org/10.1109/ACCESS.2023.3293117

Kim, J., Paik, J.-Y., & Cho, E.-S. (2023). Attention-based cross-modal CNN using non-disassembled files for malware classification. *IEEE Access*, 11, 22889–22903. https://doi.org/10.1109/ACCESS.2023.3253770

Li, S., Li, Y., Wu, X., Otaibi, S. A., & Tian, Z. (2023). Imbalanced malware family classification using multimodal fusion and weight self-learning. *IEEE Transactions on Intelligent Transportation Systems*, 24(7), 7642–7652. https://doi.org/10.1109/TITS.2022.3208891

Lu, T., & Wang, J. (2024). DOMR: Toward deep open-world malware recognition. *IEEE Transactions on Information Forensics and Security*, 19, 1455–1468. https://doi.org/10.1109/TIFS.2023.3338469

Ogala, J.O., Ahmad, S., Shakeel, I. (2023) Strengthening KMS security with advanced cryptography, machine learning, deep learning, and IoT technologies. *SN Computer Science* 4, 530. https://doi.org/10.1007/s42979-023-02073-9

Samtani, S., Chinn, R., Chen, H., & Nunamaker, J. F. (2017). Exploring emerging hacker assets and key hackers for proactive cyber threat intelligence. *Journal of Management Information Systems*, *34*(4), 1023–1053. https://doi.org/10.1080/07421222.2017.1394049

Shashi, M., Saxena, A., Vaya, D., Pillai, N., Chavan, S. S., & Balarama Krishna, C. (2023). Enhancing trust and privacy in edge and cloud computing through blockchain technology. In *2023 6th International Conference on Contemporary Computing and Informatics (IC3I)*, 1812–1816. https://doi.org/10.1109/IC3I59117.2023.10397883

Symantec by Broadcom (n.d.). Comprehensive threat protection: Integrated cybersecurity solutions. https://docs.broadcom.com/doc/icd-comprehensive-threat-protection

Vaya, D., & Hadpawat, T. (2020). Internet of everything (IoE): A new era of IoT. In *Springer: ICCCE 2019*. Lecture notes in electrical engineering 570, 1–6. https://doi.org/10.1007/978-981-13-8715-9_1

Vaya, D., & Hadpawat, T. (2021). Enhanced fingerprint authentication using blockchain. In D. Khazanchi, A. K. Vyas, K. K. Hiran, & S. Padmanaban (Eds.), *Blockchain 3.0 for sustainable development*, 123–130. De Gruyter. https://doi.org/10.1515/9783110702507-008

Wong, G.-W., Huang, Y.-T., Guo, Y.-R., Sun, Y., & Chen, M. C. (2024). Attention-based API locating for malware techniques. *IEEE Transactions on Information Forensics and Security*, 19, 1199–1212. https://doi.org/10.1109/TIFS.2023.3330337

Zhang, Y., Gui, G., & Mao, S. (2023). A lightweight malware traffic classification method based on a broad learning architecture. *IEEE Internet of Things Journal*, 10(23), 21131–21132. https://doi.org/10.1109/JIOT.2023.3297210

Zhong, F., Chen, Z., Xu, M., Zhang, G., Yu, D., & Cheng, X. (2023). Malware-on-the-brain: Illuminating malware byte codes with images for malware classification. *IEEE Transactions on Computers*, 72(2), 438–451. https://doi.org/10.1109/TC.2022.3160357

11 Safeguarding Knowledge Assets

Exploring the Nexus of Human Factors, Organizational Culture, and Social Engineering in KM Security

Manas Kumar Yogi and Dwarampudi Aiswarya

11.1 INTRODUCTION

In today's increasingly interconnected and knowledge-driven economy, organizations recognize the pivotal role of Knowledge Management (KM) in achieving sustained success. KM refers to a systematic approach to capturing, organizing, sharing, and effectively utilizing knowledge within an organization to drive performance, innovation, and competitive advantage (Guilford, 2023). As organizations seek to leverage knowledge as a core asset, they are confronted with a growing array of challenges related to the security of their knowledge assets. This introduces a critical need to examine how human factors, organizational culture, and social engineering threats intersect and influence the security of these valuable resources.

At the heart of KM are processes like knowledge creation, capture, storage, sharing, and application (Muthuswamy, 2023). Each of these processes relies heavily on the interaction between individuals and the systems in place to manage and protect knowledge. Human factors such as employee behavior, motivation, and awareness play a substantial role in either strengthening or undermining the security of knowledge assets. Negligence, insider threats, and a lack of security training make organizations vulnerable to knowledge breaches, highlighting the need for security strategies that are attuned to human behavior. Additionally, organizational culture is a key determinant in how knowledge is treated, shared, and protected. A robust security culture fosters a mindset of continuous learning, awareness, and collaboration, ensuring that knowledge assets are both safeguarded and effectively utilized. On the other hand, weak security practices within organizational culture can expose critical

 DOI: 10.1201/9781003498094-11

knowledge to significant risks, further emphasizing the need to cultivate a culture that values both KM and security (Orikpete & Ejike Ewim, 2024).

Social engineering represents another significant threat to KM security. It exploits human vulnerabilities through psychological manipulation, aiming to gain unauthorized access to knowledge assets. By understanding the tactics of social engineering, organizations can better equip employees with the necessary training and awareness to defend against such attacks. Hence, the intersection of human factors, organizational culture, and social engineering forms a nexus that must be navigated carefully to secure knowledge assets effectively in the modern organizational landscape. In this exploration, we examine how these elements—human factors, organizational culture, and social engineering—converge within the context of KM security, and how organizations can address these challenges to protect their intellectual capital and ensure sustainable growth and innovation.

11.2 SIGNIFICANCE OF SAFEGUARDING KNOWLEDGE ASSETS

Effective safeguarding of knowledge assets is crucial for organizations to maintain competitive advantage and ensure operational continuity. Protecting proprietary information, trade secrets, and innovative practices prevents competitors from gaining insights that could undermine a company's market position. Additionally, safeguarding essential data and expertise ensures uninterrupted daily operations, maintaining efficiency, productivity, and service delivery. Moreover, secure KM fosters innovation and growth by protecting intellectual property and R&D data, creating a safe environment for creativity and new developments (Prado-Gasco et al., 2015). Compliance with legal and regulatory requirements is also a key aspect, as failing to protect sensitive information can lead to severe legal penalties and damage to reputation. Financial losses from breaches or data loss can be substantial, including costs for recovery, legal actions, and compensations, along with long-term impacts like lost business opportunities. Effective safeguarding minimizes these risks, protecting against financial damage. Building trust and reputation is another critical benefit. Demonstrating a commitment to protecting sensitive information enhances relationships with customers, partners, and stakeholders, while breaches can severely damage trust and credibility. Mitigating insider threats through robust access controls, monitoring, and training ensures that only authorized individuals can access critical knowledge assets, reducing the risk of inadvertent or malicious data breaches.

Malicious insiders are employees or contractors who deliberately exploit their access to knowledge assets for personal gain or to damage the organization, while negligent insiders unintentionally compromise or expose these assets due to a lack of awareness or carelessness. Both types of insiders pose significant risks to the security of an organization's knowledge, highlighting the need for robust security measures and awareness training to mitigate these threats. Inadequate security training and insufficient awareness about the importance of KM security can lead to unintentional breaches and mishandling of sensitive information. Without proper training and understanding, employees may not follow KM security protocols effectively, increasing the risk of security vulnerabilities.

Poor communication and collaboration, often caused by organizational silos and resistance to change, can hinder the effective security of knowledge assets by isolating valuable information. Additionally, weak security policies and practices, such as inconsistent policies and a lack of accountability, create gaps in safeguarding these assets, leaving them more vulnerable to risks and breaches. Inadequate IT infrastructure, such as outdated systems and a lack of encryption, leaves KM systems vulnerable to cyber-attacks and unauthorized access. Additionally, insufficient access controls, including over-privileged access and weak authentication mechanisms, increase the risk of sensitive knowledge assets being accessed or compromised by unauthorized individuals. Phishing attacks, including deceptive emails and messages, trick employees into revealing sensitive information or providing unauthorized access to KM systems, while spear phishing targets specific individuals to access critical knowledge assets (Burton et al., 2023). Additionally, attackers use pretexting by creating false pretenses or impersonating trusted individuals or authority figures to deceive employees into divulging confidential information or granting access to sensitive knowledge.

Inadequate data classification, including the lack of proper categorization, makes it challenging to prioritize and protect critical knowledge assets, while poor management of sensitive data increases the risk of accidental exposure or loss (Nobles & McAndrew, 2023). Furthermore, insufficient backup practices can result in data loss during incidents, and ineffective disaster recovery plans can hinder an organization's ability to restore lost or compromised knowledge assets. Cyber-attacks, including hacking and malware, as well as advanced persistent threats (APTs), aim to steal sensitive knowledge assets from organizations over prolonged periods. Additionally, physical security breaches, such as unauthorized access to office spaces or data centers, can result in the theft or compromise of these assets. Natural disasters and accidents, like fires or floods, further threaten the physical integrity of critical knowledge assets, potentially causing damage or destruction.

Table 11.1 captures the interconnected roles of human factors, organizational culture, and social engineering in enhancing KM security, emphasizing the need

TABLE 11.1
Various Aspects and Their Roles in KM Security

Sl. No.	Aspect	Role in KM Security
1	Human Factors	Behavioral Aspects: Individual behaviors and decision-making processes greatly influence KM security. Employees' adherence to security protocols is crucial.
		Psychological Factors: Psychological tendencies, such as trust and authority, can be exploited by social engineering attacks. Understanding these tendencies helps in designing better security awareness programs.
		Training and Awareness: Continuous education and training programs enhance employees' awareness of security threats and proper practices, reducing the risk of human error and social engineering attacks.

(Continued)

TABLE 11.1
(Continued)

Sl. No.	Aspect	Role in KM Security
2	Organizational Culture	Security-Conscious Culture: A culture that prioritizes security fosters collective responsibility and adherence to security policies, making security an integral part of the organizational ethos.
		Leadership and Policies: Strong leadership and clear, enforced security policies ensure that security measures are implemented effectively. Leaders set the tone for the importance of security within the organization.
		Communication and Trust: Open communication and a trust-based environment encourage employees to report suspicious activities and comply with security practices, enhancing overall KM security.
3	Social Engineering	Exploitation Techniques: Social engineering attacks exploit human psychology through techniques such as phishing, pretexting, baiting, and tailgating. Understanding these methods helps in developing effective countermeasures.
		Psychological Manipulation: Attackers use principles of trust, authority, urgency, and fear to manipulate individuals into compromising security. Training programs must address these psychological manipulation techniques to reduce vulnerability.
		Real-World Scenarios: Analyzing case studies and real-world scenarios of social engineering attacks provides insights into common tactics and successful defense mechanisms, aiding in the development of robust security strategies.

for a comprehensive and adaptive approach to protect valuable knowledge assets (Campean, 2019).

Leadership plays a key role in fostering KM security by serving as role models and prioritizing resource allocation for security initiatives such as training and technology. Open communication is equally important, with transparent policies helping employees clearly understand their KM security responsibilities, while feedback mechanisms allow staff to raise concerns and suggest improvements. A collaborative environment enhances security by promoting teamwork and knowledge sharing, ensuring that critical information is protected and disseminated. Cross-functional collaboration further reinforces the consistent application of security practices and a shared understanding of KM security across the organization. Resistance to change, silos and fragmentation, and insufficient accountability undermine KM systems. Inflexibility and complacency can leave KM systems vulnerable to emerging threats, while siloed departments and isolated practices lead to poor communication and inconsistent security measures. Additionally, a lack of clear ownership and weak enforcement of policies contribute to neglect and non-compliance, further compromising KM security.

11.3 ROLE OF SOCIAL ENGINEERING IN COMPROMISING KM SYSTEMS

Table 11.2 provides a concise overview of different social engineering attack types, their mechanisms, and illustrative scenarios, highlighting the diverse tactics used by attackers to exploit human vulnerabilities and compromise security (Aldawood & Skinner, 2019).

Social engineering attacks exploit various psychological principles to manipulate individuals into compromising KM security. Understanding these manipulation techniques is crucial for developing effective defenses (Metalidou et al., 2014). Trust is a fundamental aspect of human relationships and interactions. Social engineers exploit trust to gain unauthorized access to information or systems by appearing legitimate and trustworthy. Attackers often exploit familiarity by posing as trusted individuals like colleagues, friends, or reputable companies to lower their target's defenses. They may build rapport through casual conversation or empathetic interactions to make the target more comfortable and willing to share sensitive information. Additionally, attackers might offer something of value, such as useful advice or assistance, to create a sense of obligation in the target, encouraging them to reciprocate by divulging information or granting access.

TABLE 11.2
Taxonomy of Social Engineering Attacks

Sl. No.	Type of Social Engineering Attack	Key Mechanism	Example Scenarios
1	Phishing	Deceptive emails or messages that trick recipients into revealing sensitive information or downloading malware.	An email claiming to be from IT support asks employees to reset their passwords through a provided link.
2	Spear Phishing	Targeted phishing attacks aimed at specific individuals or organizations, often using personalized information.	A CFO receives an email appearing to be from the CEO requesting an urgent transfer of funds.
3	Pretexting	Creating a fabricated scenario to obtain information or perform an action.	An attacker calls an employee, posing as a bank representative, to verify account details for security reasons.
4	Baiting	Offering a false promise of a reward to entice victims into divulging information or performing actions.	USB drives labeled as "Confidential" are left in the office, which when inserted, install malware.
5	Quid Pro Quo	Offering something in return for information or access.	An attacker offers free IT support services in exchange for login credentials.

(Continued)

TABLE 11.2
(Continued)

Sl. No.	Type of Social Engineering Attack	Key Mechanism	Example Scenarios
6	Tailgating	Gaining unauthorized access to secure areas by following an authorized person.	An attacker follows an employee into a restricted area by pretending to have forgotten their access card.
7	Vishing	Voice phishing; using phone calls to trick individuals into revealing personal information.	A caller claims to be from the IRS, threatening legal action unless the victim provides their SSN and bank details.
8	Smishing	SMS phishing; using text messages to deceive individuals into revealing sensitive information.	A text message appears to be from a bank, asking the recipient to verify their account information via a link.
9	Watering Hole Attack	Compromising a website frequently visited by the target to deliver malware.	A commonly used industry website is hacked, and malware is implanted to infect visitors from a specific company.
10	Impersonation	Pretending to be someone else to gain trust and extract information or access.	An attacker dresses as a delivery person to enter a secure facility and access sensitive areas.
11	Dumpster Diving	Searching through trash to find sensitive information.	An attacker rummages through a company's dumpster to find discarded documents with confidential information.
12	Scareware	Using alarming messages to trick victims into downloading malicious software or divulging information.	Pop-up ads claim the victim's computer is infected with viruses and prompts them to download fake antivirus software.

11.4 LITERATURE REVIEW

Understanding the psychological and behavioral aspects of employees is crucial for effective KM security. Human factors such as cognitive biases, social influence, and decision-making processes can impact security outcomes. Key psychological and behavioral aspects include (Abdullah et al., 2014):

- Cognitive Biases: Employees may exhibit cognitive biases such as confirmation bias, where they selectively interpret information to confirm pre-existing beliefs, leading to susceptibility to misinformation and manipulation.

- Trust: Employees may trust seemingly legitimate sources or requests, making them more vulnerable to social engineering attacks that exploit trust and authority.
- Complacency: Employees may become complacent or overconfident in their ability to recognize and respond to security threats, leading to a lack of vigilance and increased susceptibility to attacks.
- Fear and Anxiety: Fear of consequences or urgency to comply with perceived threats can override rational decision-making, leading employees to bypass security protocols or divulge sensitive information.
- Social Influence: Social dynamics within the organization, such as peer pressure or groupthink, can influence employees' behaviors and decisions related to security practices.

11.4.1 Culture-Driven Security Policies

Culture-driven security policies align with the organization's values, norms, and behaviors to promote a security-conscious culture and mitigate risks associated with knowledge management. Key elements of culture-driven security policies include (Nobles, 2018):

- Leadership Commitment: Senior leadership sets the tone for security culture by demonstrating commitment, support, and accountability for security initiatives. Leaders communicate the importance of security, allocating resources, and model desired behaviors.
- Clear Communication: Security policies are communicated clearly and consistently to all employees, emphasizing their role in safeguarding knowledge assets and the consequences of non-compliance.
- Employee Engagement: Employees are actively engaged in shaping security policies, providing feedback, and participating in training and awareness programs. Their involvement fosters a sense of ownership and responsibility for security outcomes.
- Recognition and Rewards: Security-conscious behaviors are recognized, rewarded, and celebrated within the organization to reinforce desired norms and encourage compliance with security policies.

11.4.2 Case Studies on Social Engineering Attacks in KM

- The Sony Pictures Hack (2014): Attackers used social engineering tactics to gain access to Sony Pictures' network and steal sensitive information, including unreleased films, executive emails, and employee data. The attack involved phishing emails targeting employees and exploiting their credentials to infiltrate the network (Nifakos et al., 2021).
- The SolarWinds Supply Chain Attack (2020): Hackers compromised SolarWinds' software update process to distribute malware to thousands of organizations, including government agencies and Fortune 500

companies (Flores et al., 2014). The attack involved social engineering tactics to gain access to SolarWinds' systems and inject malicious code into software updates.

- The Ubiquiti Networks Breach (2021): Attackers targeted Ubiquiti Networks' employees through social engineering tactics, posing as executives to obtain access credentials and infiltrate the company's systems (Lopes et al., 2022). The breach resulted in the theft of sensitive data, including source code and customer information.

11.4.3 Preventive Measures and Mitigation Strategies

- Employee Training and Awareness: Provide regular training and awareness programs to educate employees about social engineering tactics, how to recognize phishing emails, and the importance of verifying requests for sensitive information or access (Gyunka & Christiana, 2017).
- Multi-Factor Authentication (MFA): Implement MFA to require additional verification steps, such as a one-time passcode or biometric authentication, to access systems or sensitive information, even if credentials are compromised.
- Email Filtering and Anti-Phishing Tools: Deploy email filtering and anti-phishing solutions to detect and block malicious emails before they reach employees' inboxes, reducing the likelihood of successful phishing attacks.
- Access Control and Least Privilege: Implement access control policies and least privilege principles to restrict access to sensitive information and systems based on employees' roles and responsibilities, minimizing the impact of compromised accounts (Werlinger et al., 2009).
- Incident Response Planning: Develop and maintain incident response plans to effectively detect, respond to, and recover from social engineering attacks. Establish communication protocols, escalation procedures, and recovery strategies to mitigate the impact of security incidents.
- Security Awareness Campaigns: Conduct ongoing security awareness campaigns to reinforce security best practices, raise awareness of emerging threats, and promote a culture of vigilance and accountability.

11.5 ATTACK VECTORS AND SCENARIOS

In Table 11.3, different case studies which involve various cyber-attack vectors and the lessons learned from such attacks are represented (Mehdizadeh et al., 2019).

11.6 PREVENTIVE MEASURES AND COUNTERMEASURES

There has been an increase in the attack surface in the ecosystem of KM sectors, hence researchers have designed various preventive measures and counter-measures, and their effectiveness is discussed in Table 11.4.

TABLE 11.3
Taxonomy of Security Attack Vectors and Relevant Case Studies

Case Study	Overview	Methods Used	Outcomes	Lessons Learned
Target Data Breach (2013)	Theft of personal and financial information of ~40 million customers	Phishing emails to third-party vendor; Credential theft; Network access via stolen credentials	Estimated $162 million in expenses; Significant reputational damage; Regulatory actions and settlements	Secure third-party vendor access; Comprehensive employee phishing training; Network segmentation
RSA Security Breach (2011)	Compromise of RSA's SecurID two-factor authentication products	Spear-phishing emails with malicious Excel file; Malware installation; Data exfiltration	Significant remediation costs; Undermined customer confidence; Security protocol enhancements	Employee spear-phishing awareness; Robust incident response plans; Enhanced network monitoring
Google and Facebook Fraud Case (2013–2015)	Loss of over $100 million via social engineering scam	Pretexting and impersonation; Fake invoices; Manipulating employees into processing payments	Loss of over $100 million (partially recovered); Increased vigilance in payment processes; High-profile legal case	Stringent verification for financial transactions; Cross-check payment requests with known vendor contacts; Enhanced training on recognizing suspicious financial activities

11.7 FUTURE RESEARCH DIRECTIONS

In the context of KM security, researchers face several future challenges that necessitate innovative approaches and solutions. These challenges span technological advancements, evolving human behaviors, organizational dynamics, and the persistent threat landscape. As technologies such as AI, machine learning, IoT, and blockchain continue to evolve, researchers need to explore their integration into KM security frameworks. Understanding how these technologies can enhance data protection, threat detection, and incident response is crucial. Cyber threats are becoming more sophisticated, ranging from APTs to social engineering attacks. Researchers must stay ahead by anticipating and mitigating new forms of cyber threats through proactive research and development. Despite technological advancements, human factors remain a critical challenge. Understanding human behavior, decision-making processes, and perceptions of security risks is essential for developing effective training

TABLE 11.4
Preventive Measures Applicable for KM Security

Sl. No.	Category	Preventive Measure	Description	Effectiveness	Limitations/Challenges
1	Technological Defenses	Email Filtering and Anti-Phishing Tools	Detect and block phishing emails	Effective for most phishing emails; advanced detection capabilities	Sophisticated attacks may evade detection
		Multi-Factor Authentication (MFA)	Requires multiple verification methods	Increases security significantly	Can be bypassed with advanced tactics; additional factors can be compromised
		Endpoint Security Solutions	Protect devices from malware and unauthorized access	Detects and mitigates malware	Reactive; may not prevent initial social engineering attacks
		Network Segmentation	Limits spread of malware and restricts lateral movement	Reduces impact of breaches	Requires proper configuration; doesn't prevent initial attacks
2	Employee Training	Regular Security Awareness Training	Educates employees on social engineering tactics	Improves recognition and response	Effectiveness diminishes over time without reinforcement
		Phishing Simulations	Simulated phishing attacks to practice identification	Enhances employee recognition skills	May lead to desensitization; might not be taken as seriously
		Incident Response Drills	Prepares employees for responding to security incidents	Ensures familiarity with procedures	Can't cover all scenarios; drills may not replicate real stress
3	Organizational Policies	Access Control Policies	Restricts access to sensitive information	Reduces impact of compromised accounts	Overly restrictive policies may hinder productivity
		Verification Protocols	Verifies identity for access requests	Prevents unauthorized access	Time-consuming; requires strict enforcement
		Reporting Mechanisms	Facilitates reporting of suspicious activities	Enables quick response to potential threats	Fear of repercussions or skepticism may hinder reporting

(Continued)

TABLE 11.4
(Continued)

Sl. No.	Category	Preventive Measure	Description	Effectiveness	Limitations/Challenges
4	Innovative Approaches	Behavioral Analytics	Detects anomalies in user behavior	Identifies subtle signs of attacks	Requires robust data collection and analysis; risk of false positives
		Advanced Threat Intelligence	Gathers and shares information on emerging threats	Prepares for new social engineering methods	Continuous updates and integration required
		Gamified Training Programs	Uses gamification to engage employees in training	Increases participation and retention	Resource-intensive to develop and maintain
		Zero Trust Architecture	Requires continuous verification of all access requests	Continuously validates identities and access rights	Complex and costly to implement; requires mindset shift
		AI and Machine Learning	Uses AI/ML to detect and respond to attacks in real-time	Provides faster, more accurate detection	Significant investment in technology and expertise required; risk of false positives

programs and security policies. Organizational cultures vary widely and influence security practices significantly. Researchers need to explore how different cultures impact KM security and develop adaptable strategies that align with diverse organizational contexts. Compliance with data protection regulations (e.g., GDPR, CCPA) adds complexity to KM security. Researchers must navigate regulatory requirements while ensuring robust security measures without stifling innovation. Addressing KM security requires collaboration across disciplines such as cybersecurity, psychology, organizational behavior, law, and technology. Researchers need to foster interdisciplinary partnerships to tackle complex security challenges effectively.

11.8 CONCLUSION

This chapter highlights the vital role of safeguarding knowledge assets within organizations, particularly as they strive to maintain competitive advantage and operational continuity in a knowledge-driven economy. Through examining human factors, organizational culture, and social engineering, this chapter underscores how these elements can either fortify or compromise KM security. Organizations are encouraged to implement robust security measures, including employee training and awareness programs, to mitigate insider threats and reduce human error risks that could expose critical knowledge assets to malicious actors. Furthermore, organizational culture significantly influences KM security, as a strong security-conscious culture fosters collective responsibility for protecting knowledge assets. Leadership's commitment to promoting transparent communication, enforcing security policies, and building trust within teams reinforces a security-focused mindset across the organization. By fostering open communication, organizations can encourage employees to report suspicious activities and engage in security practices that uphold KM integrity.

Social engineering remains a persistent threat, exploiting human vulnerabilities through tactics like phishing, pretexting, and tailgating. This chapter emphasizes the importance of educating employees about these tactics and implementing preventive measures, such as multi-factor authentication and incident response plans. Recognizing and countering social engineering tactics enhances the organization's overall resilience, reducing the likelihood of successful attacks on KM systems. In summary, protecting knowledge assets requires a multifaceted approach that addresses human behavior, organizational culture, and social engineering vulnerabilities. Through a strategic blend of security-conscious culture, technological defenses, and continuous employee education, organizations can create a secure KM environment that supports innovation, collaboration, and sustainable growth in an interconnected world.

REFERENCES

Abdullah, H., Uli, J., & Mohamed, Z. A. (2014). Relationship between organizational characteristics and information security knowledge management implementation. *Procedia-Social and Behavioral Sciences*, 123, 433–443. https://doi.org/10.1016/j.sbspro.2014.01.1442

Aldawood, H., & Skinner, G. (2019). Reviewing cyber security social engineering training and awareness programs—Pitfalls and ongoing issues. *Future Internet*, 11(3), 73. https://doi.org/10.3390/fi11030073

Burton, S. L., Nobles, C., Zurek, J., Balos, D., & Mutton, N. (2023). Exploring the nexus of cybersecurity leadership, human factors, emotional intelligence, innovative work behavior, and critical leadership traits. *Scientific Bulletin*, 28(2), 162–175.

Campean, S. (2019). The human factor at the center of a cyber security culture. *International Journal of Information Security and Cybercrime (IJISC)*, 8(1), 51–58.

Flores, W. R., Antonsen, E., & Ekstedt, M. (2014). Information security knowledge sharing in organizations: Investigating the effect of behavioral information security governance and national culture. *Computers & Security*, 43, 90–110. https://doi.org/10.1016/j.cose.2014.03.004

Guilford, M. (2023). Systemic risk analysis of human factors in phishing. (Doctoral dissertation, Old Dominion University).

Gyunka, B. A., & Christiana, A. O. (2017). Analysis of human factors in cyber security: A case study of anonymous attack on HBGary. *Computing & Information Systems*, 21(2), 17–25.

Lopes, A., Sousa, P., Almeida, C., & Silva, A. (2022). Information security threat assessment using social engineering in the organizational context–Literature review. In *World conference on information systems and technologies*. Springer International Publishing.

Mehdizadeh, P., Chizari, M., & Nouraei, H. (2019). Investigating the underlying factors of the establishment of knowledge management in insurance organizations: Case study in social security organization. *Journal of Military Medicine*, 21(4), 353–361.

Metalidou, E., Marinagi, C., Trivellas, P., Eberhagen, N., Skourlas, C., & Giannakopoulos, G. (2014). The human factor of information security: Unintentional damage perspective. *Procedia-Social and Behavioral Sciences*, 147, 424–428. https://doi.org/10.1016/j.sbspro.2014.07.141

Muthuswamy, V. V. (2023). An empirical exploration of human factors, sense of security, and well-being in contemporary social work environments. *Journal of Human Security*, 19(2), 343–356.

Nifakos, S., Stergiopoulos, S., Katsikas, S., & Gritzalis, D. (2021). Influence of human factors on cyber security within healthcare organisations: A systematic review. *Sensors*, 21(15), 5119. https://doi.org/10.3390/s21155119

Nobles, C. (2018). Botching human factors in cybersecurity in business organizations. *HOLISTICA–Journal of Business and Public Administration*, 9(3), 71–88.

Nobles, C., & McAndrew, I. (2023). The intersectionality of offensive cybersecurity and human factors: A position paper. *Scientific Bulletin*, 28(2), 215–233.

Orikpete, O. F., & Ejike Ewim, D. R. (2024). Interplay of human factors and safety culture in nuclear safety for enhanced organizational and individual performance: A comprehensive review. *Nuclear Engineering and Design*, 416, 112797.

Prado-Gasco, V. J., Ros, S. M., & Navarro, C. (2015). Knowledge management in R&D teams at a Spanish technical university: Measurement and relations with organizational culture. *Inžinerinė ekonomika*, 26(4), 398–408.

Werlinger, R., Hawkey, K., & Beznosov, K. (2009). An integrated view of human, organizational, and technological challenges of IT security management. *Information Management & Computer Security*, 17(1), 4–19. https://doi.org/10.1108/09685220910944722

12 The Role of Knowledge Management in Cybersecurity for Digital Marketing Campaigns

Rajasekhara Mouly Potluri, Mukashev Erzhan Bulatovich, and Lohith Sekhar Potluri

12.1 INTRODUCTION

Current technological advances have accelerated the globalization process in all business sectors. The rapid spread of information directly impacts the global economy, making it essential for all technology-adaptive countries to consider the economic influence on each sector (Nugroho et al., 2021). In this digitalized world, knowledge management is pivotal in the corporate world due to the extensive application of both soft and hard copies of knowledge or information in all operative functions of management. Poston (2019) emphasized that data has been called "the new oil," and it is crucial for real people to access all the company's data types to gain a competitive advantage in the current highly competitive business environment. Knowledge management, under the leadership of the knowledge manager, who manages all aspects of the organization's intellectual resources, is imperative in every company. The unique task of managing and using data by only authorized people from the concerned departments should be accessible to those who perform their duties, and all data should be appropriately secured within the company. Throughout the globe, extensive data thefts have occurred irrespective of the type of business and the company's magnitude. With this kind of backdrop, every organization should carefully develop a protective cybersecurity system for every department or functional activity of the current situation. Most organizations around the world treat cybersecurity as a crucial knowledge environment, which is helpful in attracting talented people. Administering cybersecurity is a pivotal element in managing capabilities and improving domain-oriented talent discovery. In today's digital era, cybersecurity knowledge management is not just a luxury but an obligation for organizations seeming to shield their wealth and reputations. As the cybersecurity topography persistently changes, businesses face the enormous experiment of staying reorganized on emerging threats and best procedures. Competent cybersecurity knowledge management systems serve as an organization's first line of defense, confirming that

DOI: 10.1201/9781003498094-12

information about the latest liabilities, attack vectors, and mitigation techniques is centralized, current, and accessible.

By consolidating this vital data, organizations can ensure that their cybersecurity teams are well-equipped to respond swiftly and effectively to threats. Applicable cybersecurity knowledge management extends beyond just packing information; it is about establishing a dynamic environment where information emerges flawlessly, encouraging teams to conduct knowledge against potential cyber threats. With hackers acquiring more innovative and new security threats cracking up every day, companies must proactively thwart occurrences before they happen. There are many ways to do this, and the company focuses on knowledge sharing. Companies' functional departments are all connected, so their security efforts are also connected. If hackers find vulnerabilities in one system, they will likely see and exploit them in other systems within the same community. With data sharing at the core of the cloud software industry, an attack on one system can quickly become an attack on many. Companies improve resilience by sharing knowledge, experiences, and updates about cybersecurity threats with the broader community, maintaining a more robust security program, and helping others keep theirs top-notch.

12.2 THE SIGNIFICANCE OF KNOWLEDGE MANAGEMENT IN CYBERSECURITY

Identifying the substantial discrepancy between technology and knowledge within an organization is overwhelming. Companies are spending significant time and resources to convert their yesteryear's artificial organization into a technology-based organization by introducing competent knowledge creation, organization, and storage. Girdhar highlighted in his blog why knowledge management is fundamental for a successful information security program and enunciated how the Building a Second Brand (BASB) approach, which includes the CODE and PARA methods, can increase the organization's overall security. In information security, Personal Knowledge Management (PKM) ensures that the workforce can access the relevant and up-to-date knowledge needed to prevent breaches and protect critical data. Mr. Girdhar emphasized that the Building a Second Brain concept, popularized by Tiago Forte, proffers an innovative approach to PKM, which lies on the CODE method (Capture, Organize, Distill, Express) combined with the PARA method (Projects, Areas, Resources, Archives), which enables organizations to create a robust information security program.

These methods stressed the diverse advantages such as improved awareness and decision-making through capturing and distilling crucial information, enhanced collaboration with information sharing with the express, streamlined processes, reduced redundancy using the PARA method, ongoing learning, and adaptation through continuous review, enhanced employee confidence and competence with organized knowledge. However, effective knowledge management is the basis for protecting the organization's all kinds of information; the aggressiveness of scammers is attacking continuously, throwing sleepless nights at the think tanks of many companies in different parts of the world. Cyber security (CS) safeguards information and technology assets that support and rely on cyberspace, users, and cyberspace itself from cyber-attacks (Von Solms & Van Niekerk, 2013). Even though there are some dilemmas in administering cybersecurity, it is a must situation for all organizations because of the extensive perils identified irrespective of the type of sector they are operating, like public, private, critical infrastructure, and surprisingly, including military of many countries. These predicaments first involve balancing the interests of private, public, and infrastructure projects. The second is data protection and privacy against information and knowledge sharing (Klimburg, 2012). Furthermore, recruiting, improving, and retaining competent people are essential issues for cybersecurity. These cybersecurity-knowledgeable people are periodically stated to be talented (Caldwell, 2013; Conrad, 2012).

A Knowledge Management (KM) system in cybersecurity is a tool that addresses cybersecurity issues by considering the social, organizational, and technological factors involved. It involves the analysis of constructs such as the roles of KM in cybersecurity, organizational framework, cybersecurity analytics process, tools, and system architecture. This framework provides guidelines for developing KM systems for cybersecurity in organizations, especially small and medium-sized enterprises, in their information and cybersecurity risk management. Information and cyber security risk management establish knowledge-intensive business processes that can be assisted by a proper KM system (Nykänen & Kärkkäinen, 2018). Another approach is using Master Data Management (MDM) technology to develop a state system segment for detecting, preventing, and eliminating cyber-attack consequences. KM is an instrument to tackle cybersecurity issues, and it stresses the connected social, organizational, and technological factors implicated in cybersecurity. This paper proposes a sociotechnical systems analysis framework of KM systems for cybersecurity. Specifically, it applies to a sociotechnical systems approach to investigating the constructions of KM systems for cybersecurity. It identifies five significant constructs of KM systems for cybersecurity: roles of KM in cybersecurity, organizational framework of KM for cybersecurity, cybersecurity analytics process, tools of KM for cybersecurity, and system architecture of KM for cybersecurity. (Wang & Wang, 2021). Additionally, Semantic MediaWiki can be used as a platform to assist organizations, as well as tiny and medium-sized enterprises, in their information and cyber security risk management (Petrenko et al., 2018). KM plays a prominent role in cybersecurity. This study from Wang and Wang (2019) garners five real-life cases of best practices of KM for the domain of cybersecurity in business organizations. Through an iterative procedure of team-based qualitative data analysis of the five cases, the study develops a model of knowledge management for cybersecurity that conceptualizes three common aspects of KM practices for cybersecurity in business organizations. First, KM for cybersecurity in business organizations has evident, specified organizational structures that involve three inter-organizational layers among the organization frontiers. Second, the knowledge flows of KM for cybersecurity in business organizations underline precise, indicative, and identifiable knowledge. Third, in contrast to KM for other domains, KM for cybersecurity has well-defined objective measures to evaluate the efficacy of KM. The domain-specific KM model grounded on outstanding practice cases proffers a road map for KM procedures in the cybersecurity sphere in business organizations.

12.3 THE ROLE OF DIGITAL MARKETING IN BUSINESS PERFORMANCE

Digital marketing is growing increasingly popular and is recognized as one of the most successful strategies for driving sales, particularly for small enterprises, start-up ventures, and quick access to overseas markets. The global crisis has impacted the economies of numerous countries, including Kazakhstan. Due to severe external crisis conditions and fierce rivalry, companies are required to be business-active in the market. Internet marketing is a significant promotional tool for Kazakhstan's economy, both in the domestic market and on a global scale. Digital marketing is defined as marketing that involves engagement with clients and business partners using digital information and communication technology and electronic devices (Stavrova et al., 2021). Digital marketing includes Internet marketing, but it allows businesses to communicate with potential consumers not only online but also in an offline environment (outdoor advertising displays, QR codes, mobile applications, SMS/MMS, video walls, interactive performances). Through these forms of advertising, the target audience penetrates the online environment through an offline environment (Appel et al., 2020).

For the first time, the term "digital marketing" was mentioned in 1990 when the world's first search engine "Archie" was launched with the possibility of search engine optimization (raising the position of the site for specific queries). The digital era began with the advent of the Internet and the development of the Web 1.0 platform (Crittenden & Crittenden, 2015). The Web 1.0 platform allowed users to find the information they needed but did not allow them to share this information over the Internet and discuss it. In the 1990s, CRM software (Customer Relationship Management System) became very popular and transformed into eCRM after the advent of the Internet. In 1993, the first clickable banner appeared on the Internet, after which AT&T conducted the first digital advertising campaign, in which 44% of consumers who saw the banner clicked on it (Bala & Verma, 2018). Today, digital marketing is one of the most effective elements of the marketing communications complex, and it shows no signs of slowing down or stopping. Therefore, leading companies and their marketers focus their efforts on forming digital communications with their consumers (Kane, 2015). However, at the same time, digital marketing has its specific features, such as continuity in time and space, because the Internet operates without a time limit (24 hours a day, 365 days a year); therefore, any marketing campaign that is carried out on the Internet operates without stopping in time; versatility—the exact implementation of network protocols for information exchange (HTTP, SMTP, FTP, etc.) anywhere in the world, which leads to the absence of restrictions for the use of the same marketing campaign on a global scale (Strauss et al., 2006);—multichannel—there are many channels for product promotion in Internet marketing. These are websites, mobile versions of websites, mobile applications, mobile communication tools, mobile social networks, and applications (Zhao, 2005).

The central systems of interaction in Kazakhstan are: 1) B2B (Business-to-business)—business-to-business, which implies the sale of goods or services from one company to another. The main ways of interaction between market participants are electronic vertical markets—a trading community in a particular industry; B2B communities—trading platforms of suppliers; B2B-Web storefronts; B2B-auctions—trading platforms for the exchange of offers between suppliers and buyers of industrial goods; B2B—electronic exchanges (in the Republic of Kazakhstan - KASE)—trading platforms for automated selection of sellers and buyers of industrial goods; intersectoral B2B portals—trading platforms that unite the markets of various sellers and buyers. 2) B2C (Business-to-consumer)—business to the end consumer, that is, selling goods directly to the end consumer (retail). The main ways of interaction between market participants are online retail sales of services in real-time, such as transport, education, tourism, financial, etc.; B2C Web storefronts; and online catalogs (Lund & Manyika, 2016). 3) G2E (Government-to-everyone) or (E-government)—electronic government, that is, the provision of services by public authorities to individuals and legal entities persons. The main ways of interaction are portals of the government and other public authorities, where you can get information, use interactive requests, and get the necessary services. 4) C2B (Consumer-to-business)—consumer-to-business, that is, the consumer offers enterprises their price. The main ways of interaction are reverse electronic auctions (Turkyilmaz et al., 2021).

Now, the authors are analyzing the development of Internet marketing in the Republic of Kazakhstan. The Internet first appeared in Kazakhstan in 1993. Since then, the rapid growth of the Internet in Kazakhstan has begun. In 2000, 70 thousand Internet users were registered in the Republic, and by 2012, the number of Internet users had increased to 9.4 million, that is, 134 times (Gommans et al., 2021). In 2013, the number of people in the Republic of Kazakhstan using the Internet increased to 65.6% of the Republic's total population. Currently, e-commerce is actively growing in Kazakhstan, and according to STZ forecasts, by the beginning of 2026, the e-commerce market volume should be more than $4.6 billion (Tovma & Ploshai, 2015). In 2018, according to various estimates, the digital market in Kazakhstan was estimated at $1.6–1.8 billion, which was 0.7% of the total share in retail. Thus, according to the Johnson & Partners consulting agency, in 2018, the e-commerce market in Russia was estimated at $15 billion (2% share in retail), in Ukraine—$2.2 billion (2%), and in Belarus—$350 million (1.2%). Thus, the Internet commerce market in Kazakhstan is still underdeveloped and is at the implementation stage (O'Brien et al., 2019). In the e-commerce segment, there are about 7500 companies in Kazakhstan. According to Forbes Kazakhstan, the leading players in Internet commerce are JSC Air Astana (net revenue—about $80 million), AlserComputers LLP ($6.6 million), Kazkommertsbank JSC ($3.3 million) and Arena S LLP Sulpak.kz ($6.3 million). Digital marketing in the Republic of Kazakhstan is in its infancy. However, the most progressive-minded business people are already actively positioned on the Internet, using videos, blogs, and social media accounts to communicate with consumers and organize online promotional tours. The leading Internet sites in the Republic of Kazakhstan are Google, YouTube, Instagram, and TikTok. The main sales tools of Internet marketing in Kazakhstan are contextual advertising in search results, email marketing, search engine optimization, and targeted advertising on social networks. The primary tool for forming customer loyalty in Internet marketing Kazakhstan's goal is to promote on social networks. According to Kazmedia experts, the volume of the online advertising market in 2021 amounted to about $2 billion. Yandex earned in 2020 about $1.4 billion a year in the Kazakh market, and this is not the limit but only the beginning of the development of online advertising in Kazakhstan (Zhekenov et al., 2020).

The significant local electronic resources in Kazakhstan for Internet marketing are vse.kz, yvision.kz, kolesa.kz et al. Regarding return on investment, the most effective Internet marketing tools are promotion tools such as SEO optimization, contextual advertising, and email marketing (Raimbekov et al., 2018). SEO or search engine optimization is a whole sequence of specific actions performed on a site to increase its position on the search results pages by search engine systems for overall performance improvement, improvement, and restructuring of the site's activities. Thus, the strengths of the development of Internet marketing in the Republic of Kazakhstan are the rapid growth of Internet consumers, as well as the main activities of Internet marketing in Kazakhstan; the availability of an adequate legislative framework and government support; the development of Internet experience by consumers in the Republic of Kazakhstan, which leads, respectively, to an increase in Internet access; the presence of the first successful companies specializing in the provision of Internet marketing services; the openness of Kazakhstan's economy has led to the emergence of the most prominent players in the country's Internet market; the rapid pace of formation telecommunications infrastructure in the Republic of Kazakhstan; availability of ample local electronic resources in Kazakhstan for Internet marketing. The weaknesses of the development of Internet marketing in the Republic of Kazakhstan are: consumer confidence in Internet sales is insufficiently formed; cash payment for most online purchases due to the lack of Internet access to most bank cards, which indicates a weak development of convenient means of payment; weak development of inexpensive and convenient infrastructure for the delivery of goods; insufficient organization of SEO optimization domestic companies, which in the current market conditions is the most suitable tool for promoting Internet resources; weak activity of local electronic resources (Chung, 2019).

Based on the above, the effectiveness of digital marketing directly depends on how interconnected the tools used are and to what extent their actions support each other. The main advantages of digital marketing: interaction with a wide range of target audiences; the opportunity to inform a wide range of the public in a short time; working with clients through feedback, which ensures that all information is received; greater user engagement through powerful influence on them; increasing brand credibility and showing trust in it; increasing competitiveness through a non-standard approach; the ability to monitor the results in real-time; attracting new customers and increasing sales.

Digital marketing also has significant disadvantages: comprehensive study of customized advertising; slow development of specific sectors of the economy; general low level of market development, including imperfect competition, low level of solvency of the population, the heavy introduction of innovations into existing economic processes, etc.; opacity of market data, that is, inconsistency of statistical data provided about the economy with the accurate picture; inconstancy of market research; the disparity of data obtained by analysts in the analysis of the market; – lack of personnel in the marketing field; the need for advertising. Due to the intensive dynamics of the development of the modern market, companies must use new, contemporary digital marketing tools to achieve success. Digital marketing has a high social and communication effect, significantly increasing the economic efficiency of the company's activities. Moreover, it forms an image of the brand of the product and the desire to purchase it. It also makes it possible to accelerate the process of attracting new customers, effectively promote services and products, and the ability to send personalized content. Thus, digital marketing in Kazakhstan's business performance has developed quickly with the advent of modern information technologies. Its influence has increased, and it can be assumed that this area will continue to grow as a type of business and a source of information dissemination.

12.4 CYBERSECURITY IN DIGITAL MARKETING

In an ever-expanding digital landscape, digital marketing is the world's most exciting marketing communication mix element. It competently offers a tremendous competitive advantage to the company with its most accurate and influential promotion. The company's digital marketing efforts have confidently influenced the buying behavior of all classes of consumers throughout the globe because of its unmatched features and extensive use of social media. In light of these influential marks on the minds of target market consumers, it is imperative situation to the corporate world to protect and safeguard their digital marketing actions and strategies against cyber threats to maintain trust among the consumers. Gadhavi (2024) emphasized the significance of cybersecurity in fortifying companies' digital marketing strategies against potential breaches, along with measures and steps to shield their online presence. As Mr. Gadhavi stressed, cyber threats are persistently increasing in almost all markets, including developed markets; developing cohesiveness on cybersecurity fundamentals is not only a proactive action but also an integral aspect of ensuring the long-term integrity and success of the company's digital marketing initiatives.

The significance of cybersecurity in digital marketing is paramount due to the extensive observation of cyber-attacks from different fronts. Protecting sensitive company data and maintaining faith and confidence are critical for success. Cyber-attacks can lead to reputational damage, financial losses, and legal consequences. It is a must situation to implement robust security measures, such as encryption technology, regular system updates, and employee training to protect the company's all kinds of data from cyber-attacks. Suppose any company has shown lenience in protecting its data. In that case, they will face severe consequences from the control mechanisms of their respective countries, along with stagnating their company's growth and success. In the spear of digital marketing, cybersecurity plays a pivotal role in safeguarding businesses, customers, and other sensitive data concerned with operative functions of management, viz., finance, human resources, and operations. These ever-growing cyber-attacks always have the capacity to create notable perils beyond the protection measures introduced by the corporate world and throw challenges in safeguarding company websites, social media accounts, and online promotions.

In today's highly competitive business world, cybersecurity is a commanding tool and technique for shielding a company's digital marketing activity and protecting all kinds of data. Concerned with the digital marketing field, it helps to protect brand reputation and customer fidelity. The cybersecurity team of a company or concern develops a protective wall and acts as an impediment against potential threats to safeguard a digital marketing milieu with the application of technology and the Internet. A betrayal that reveals customer data can substantially affect their confidentiality, leading to reputational impairment and financial losses; identity theft and personal data leaks are two of the most common threats in the digital world, making it essential for businesses to protect their customer's confidential information. Digital marketing, with its intricate web of online platforms and tools, is a prime target for cyber-attacks. A hacked website or compromised online presence can disrupt operations, eliminate trust, and damage the brand's reputation. Businesses must take cybersecurity seriously to navigate this complex digital landscape safely, if possible, with their resource base. Alternatively, collaborating with reputable cybersecurity firms is a strategic choice (EC-Council University, 2024).

12.5 SIGNIFICANT CYBERSECURITY THREATS AND SOLUTIONS TO PROTECT DIGITAL MARKETING

The most influential promotional mix element in digital marketing in the present digitalized corporate world, the most common cybersecurity threats identified are phishing attacks, malware, ransomware, data breaches, and distributed denial of service (DDoS). Gadhavi (2024) suggested excellent and careful practices for cybersecurity in digital marketing. The corporate world has to protect its digital marketing assets against cyber-attacks through strong passwords and two-factor authentication, regular software updates and patches, employee training and awareness, secure data storage, and encryption. Focusing on cyberspace security troubles is critical to triumph in digital marketing, which is an online business. As the digital age evolves, more people select to work online and start digital and technological enterprises. Marketing is a well-known good example. Digital marketing is a novel marketing genre that has emerged in the new millennium. Digital marketing is a profitable and crucial industry since it can benefit almost any organization. Here are some typical forms of cyber-attacks that imply digital marketing: Malware infection from files downloaded or links clicked; Browser hijacking and redirection; Stealing of data and other sensitive information; Identity theft; Proliferation of fake news; DDoS attacks on the website; WordPress malware (Kumar et al., 2022).

Critical cybersecurity threats in digital marketing, including phishing attacks, malware and ransomware, data breaches, and DDoS attacks, are explored, emphasizing their potential impact on business operations, customer trust, and brand reputation (Şenyapar, 2024). In digital marketing, cybersecurity plays a pivotal role (Agarwal, 2021; Das, 2021; Sisodia & Sisodia, 2023). Cybersecurity in digital marketing refers to the comprehensive set of practices, strategies, and technologies employed to protect digital marketing assets from a wide range of cyber threats and attacks (Huzaizi et al., 2021). Since digital marketing frequently involves collecting, storing, and processing sensitive customer data, including personal information and payment details, it becomes a prime target for cybercriminals (Dubovyk et al., 2022; Konyeha, 2020). These cyber threats can take numerous forms, including hacking attempts, phishing scams, malware attacks, and data breaches (Garg et al., 2022; Konyeha, 2020; Rosário, 2023). In this case, cybersecurity measures are critical to safeguard data privacy and ensure marketing campaigns' integrity and continuity. These measures include the implementation of robust encryption protocols, regular security audits, using secure and updated software, and adopting stringent data protection policies (Alawida et al., 2022).

Moreover, effective cybersecurity in digital marketing is not only about protecting against financial and data losses. It plays a crucial role in maintaining customer trust and loyalty, which are integral to the success of any business (Iyas et al., 2021). In an era where consumers are increasingly aware of data privacy concerns, demonstrating a commitment to cybersecurity can significantly enhance a brand's reputation and customer relationships (Krishen et al., 2021). In light of the extensive usage of digital marketing campaigns by the present day's business world, irrespective of their size and type, their campaigns are extensively vulnerable to cyber threats from hackers that can compromise their efficacy and veracity. The sophistication of these attacks has grown, with cybercriminals using advanced tactics like social engineering to make their attempts more convincing, thereby increasing the risk of compromised sensitive information (Konyeha, 2020; Srivastav & Gupta, 2021). By implementing competent protective measures, businesses worldwide persistently introduce different techniques to safeguard their digital marketing campaigns from malicious threats. These techniques include implementing a secure firewall, using safe and encrypted communication channels, monitoring and analyzing network traffic, and regularly backing up and updating data.

12.6 CONCLUSION

Digital marketing is a powerful tool for companies to reach their target audience, even if they face all the risks of cyber-attacks. These risks have been steadily enhancing as per Verizon's 2024 Data Breach Investigations Report, which states a 180% rise in attacks this year. Most focused on using vulnerabilities in digital platforms; the main entry points were web applications (Shaikh, 2024). In this extensive spread of cyber-attacks, the corporate world, from small and medium enterprises to multinationals, has compulsorily and massively used digital marketing to introduce KM to their cybersecurity system. The intersection of cybersecurity and digital marketing is a crucial aspect of current business strategy. As digital marketing continues to develop, specifically in the present 21st century, integrating robust cybersecurity measures with the support of developing substantial KM is not just an option but a necessity for protecting data, maintaining customer integrity, and ensuring compliance with the respective country's legal standards. The future of digital marketing entirely depends on perfectly backed-up cybersecurity measures, which is of utmost significance as technology continues to move forward. With the backdrop of KM, companies must prioritize cybersecurity with modern and sophisticated technological tools and techniques to protect their digital assets competently, foster customer loyalty, and maintain a competitive advantage in the digital marketplace.

REFERENCES

Agarwal, M. (2021). A study on pay-per-click advertising. *Asian Journal of Multidimensional Research, 10*(11), 618–624.

Alawida, M., Omolara, A. E., Abiodun, O. I., & Al-Rajab, M. (2022). A deeper look into cybersecurity issues in the wake of Covid-19: A survey. *Journal of King Saud University - Computer and Information Sciences*, 34(10, Part A), 8176–8206. https://doi.org/10.1016/j.jksuci.2022.08.003

Appel, G., Grewal, L., Hadi, R., & Stephen, A. T. (2020). The future of social media in marketing. *Journal of the Academy of Marketing Science*, *48*(1), 79–95.

Bala, M., & Verma, D. (2018). *A Critical Review of Digital Marketing. International Journal of Management, IT & Engineering*, *8*(10), 321–339.

Caldwell, T. (2013). Plugging the cyber-security skills gap. *Computer Fraud & Security, 2013*(7), 5–10.

Chung, K. C. (2019). Mobile (shopping) commerce intention in Central Asia: The impact of culture, innovation characteristics and concerns about order fulfillment. *Asia-Pacific Journal of Business Administration*, *11*(3), 251–266.

Conrad, J. (2012). Seeking help: The important role of ethical hackers. *Network Security, 2012*(8), 5–8.

Crittenden, V., & Crittenden, W. (2015). Digital and social media marketing in business education: Implications for the marketing curriculum. *Journal of Marketing Education*, *37*(2), 71–75.

Das, S. (2021). *Search Engine Optimization and Marketing: A Recipe for Success in Digital Marketing*. CRC Press, A Chapman and Hall Book.

Dubovyk, T., Buchatska, I., Zerkal, A., & Lebedchenko, V. (2022). Digital marketing in the condition of wartime posture in Ukraine, *International Journal of Computer Science & Network Security*, *22*(7), 206. https://doi.org/10.22937/IJCSNS.2022.22.7.25

EC-Council University (2024). The importance of cybersecurity in digital marketing. https://www.eccu.edu/blog/cybersecurity/the-importance-of-cybersecurity-in-digital-marketing/

Gadhavi, J. (2024). Learn to safeguard your digital marketing with cyber security. *Sun Media Marketing*. https://www.sunmediamarketing.com/digital-marketing/digital-marketing-cyber-security/#

Garg, S., Gupta, S., & Gupta, B. (2022). Impacts of blockchain on digital marketing. In *Intelligent Sustainable Systems: Selected Papers of WorldS4 2021, Volume 1* (pp. 209–217). Springer Singapore.

Gommans, M., Krishnan, K.S., & Scheffold, K.B. (2021). From brand loyalty to e-loyalty: A conceptual framework. *Journal of Economic and Social Research*, 3(1), 43–58.

Huzaizi, A. H. A., Tajuddin, S., Bahari, K. A., Manan, K. A., & Abd Mubin, N. N. (2021). Cyber-security culture towards digital marketing communications among small and medium-sized (SME) entrepreneurs. *Asian Culture and History*, 13(2), 1–20.

Iyas, G. B., Munir, A. R., Tamsah, H., Mustafa, H., & Yusriadi, Y. (2021). The influence of digital marketing and customer perceived value through customer satisfaction on customer loyalty. *Journal of Legal, Ethical and Regulatory Issues*, *24*(Pt. 2), 1.

Kane, G. C. (2015). *Strategy, not technology, drives digital transformation.* MIT Sloan Management Review and Deloitte University Press.

Klimburg, A. (2012). *National cyber security framework manual.* NATO CCD COE Publication.

Konyeha, S. (2020). Exploring cybersecurity threats in digital marketing. *NIPES-Journal of Science and Technology Research*, *2*(3), 445–458.

Krishen, A. S., Dwivedi, Y. K., Bindu, N., & Kumar, K. S. (2021). A broad overview of interactive digital marketing: A bibliometric network analysis. *Journal of Business Research*, 131, 183–195.

Kumar, S., Pallathadka, H., & Pallathadka, L. K. (2022). An analysis of cyber security threats in digital marketing. *Journal of Critical Reviews*, *9*(03), 85–94.

Lund, S., & Manyika, J. (2016). *How digital trade is transforming globalization. E15Initiative*. International Centre for Trade and Sustainable Development (ICTSD) and World Economic Forum.

Nugroho, A., Takahashi, M., & Masaya, I. (2021). Village fund asymmetric information in disaster management: Evidence from village level in Banda Aceh City. In *IOP Conference Series: Earth and Environmental Science* (Vol. 630, No. 1, p. 012011). IOP Publishing.

Nykänen, R., & Kärkkäinen, T. (2018). A knowledge interface system for information and cyber security using semantic Wiki. In *Designing for a Digital and Globalized World: 13th International Conference, DESRIST 2018*, Chennai, India, June 3–6, 2018, Proceedings 13 (pp. 316–330). Springer International Publishing.

O'Brien, D., Main, A., Kounkel, S., & Stephan, A. (2019). 2020 global marketing trends: Bring authenticity to our digital age. https://www2.deloitte.com/content/dam/insights/us/articles/2020-global-marketing-trends/DI_2020%20Global%20Marketing%20Trends.pdf

Petrenko, S., Makoveichuk, K., & Olifirov, A. (2018, November). New methods of the cybersecurity knowledge management analytics. In *International Conference on Convergent Cognitive Information Technologies* (pp. 296–310). Springer International Publishing.

Poston, H. (2019). Knowledge management in cybersecurity. https://www.infosecinstitute.com/resources/general-security/knowledge-management-in-cybersecurity/

Raimbekov, Z., Syzdykbayeva, B., Rakhmetulina, Z., & Zhenskhan, D. (2018). The effectiveness of logistics development and its impact on the economies of the countries along the silk road passing through Kazakhstan. *Transport Problems*, *13*, 387–395.

Rosário, A. T. (2023). Security in digital marketing: Challenges and opportunities. In *Confronting security and privacy challenges in digital marketing* (pp. 206–233). IGI Global. https://doi.org/10.4018/978-1-6684-8958-1.ch010

Şenyapar, H. N. D. (2024). Digital marketing in the age of cyber threats: A comprehensive guide to cybersecurity practices. *The Journal of Social Science*, *8*(15), 1–10.

Shaikh, A. (2024). Exploring the cyber security risks of digital marketing. https://itsecuritywire.com/featured/exploring-the-cyber-security-risks-of-digital-marketing/

Sisodia, D., & Sisodia, D. S. (2023). A transfer learning framework towards identifying fraudulent publishers' behavioural changes in pay-per-click online advertising model for click fraud detection. *Expert Systems with Applications*, 232, 120922.

Srivastav, P., & Gupta, H. (2021, September). Role and Applications of digital marketing in digital era: A review. In *2021 9th International Conference on Reliability, Infocom Technologies and Optimization (Trends and Future Directions) (ICRITO)* (pp. 1–5). IEEE.

Stavrova, E., Zlateva, D., & Pinelova, L. (2021). The digital transformation in the service of business. *Economics and Management*, *18*(1), 128–136.

Strauss, J., El-Ansary, A., & Frost, R. (2006). *E-Marketing* (4th ed). Pearson Prentice Hall.

Tovma, N.A., & Ploshai, A.G. (2015). Analysis of trends in the development of internet marketing in Kazakhstan. *Bulletin of KazNU*, 12, 56–62.

Turkyilmaz, A., Dikhanbayeva, D., Suleiman, Z., Shaikholla, S., & Shehab, E. (2021). Industry 4.0: Challenges and opportunities for Kazakhstan SMEs. *Procedia Cirp*, *96*, 213–218.

Von Solms, R., & Van Niekerk, J. (2013). From information security to cyber security. *Computers & Security*, 38, 97–102.

Wang, S., & Wang, H. (2019). Knowledge management for cybersecurity in business organizations: A case study. *Journal of Computer Information Systems*, *23*, 78–93.

Wang, S., & Wang, H. (2021). A sociotechnical systems analysis of knowledge management for cybersecurity. *International Journal of Sociotechnology and Knowledge Development (IJSKD)*, *13*(3), 77–94.

Zhao, F. (2005). Entrepreneurship and innovation in e-business: An integrative perspective. *The International Journal of Entrepreneurship and Innovation*, *6*(1), 53–60.

Zhekenov, D., Darkenbaev, A., & Zhunisov, Z. (2020). The'Nurly Zhol'and the'One Belt, One Road'projects: Juxtaposition of an idea. In *The challenge of change for the legal and political systems of Eurasia: The impact of the new silk road*, Al-Farabi Kazakh National University (pp. 131–140).

13 Knowledge Management and Cybersecurity in Higher Educational Institutions After COVID-19

A Case of Kazakhstan

Rajasekhara Mouly Potluri, Turarbekova Kuralay, and Junnussov Aruan

13.1 INTRODUCTION

Knowledge management is a critical area of focus in contemporary organizations, recognizing the value of effectively leveraging intellectual assets for sustainable competitive advantage. This study delves into the intricacies of knowledge management, exploring its implications and applications within diverse organizational contexts. The catastrophe triggered by the array of COVID-19 coronavirus has affected the global economic, political, and corporate landscapes. The COVID-19 pandemic has undeniably reshaped the fabric of our global society, causing unprecedented challenges across various spheres (Arcia et al., 2021). As we gradually emerge from the immediate crisis response phase, attention turns toward the aftermath and the need for strategic approaches to navigate the new normal. Among the critical aspects demanding attention is knowledge management—the structured process of creating, sharing, and utilizing knowledge within organizations. The profound disruptions caused by the pandemic have compelled businesses, institutions, and governments to reevaluate their knowledge management strategies. The lessons learned during this crisis underscore the importance of resilience, adaptability, and efficient knowledge transfer in the face of unforeseen challenges. In the post-pandemic landscape, the effectiveness of knowledge management will play a focal role in revealing an organization's ability to innovate, recover, and thrive. This essay delves into the evolving dynamics of knowledge management in the wake of the pandemic. The researchers explore the challenges organizations face, the innovations spurred by necessity, and the long-term implications for how knowledge is generated, shared, and utilized in a world

DOI: 10.1201/9781003498094-13

forever changed by the events of COVID-19. Understanding and optimizing knowledge management processes becomes a strategic advantage for building a resilient and forward-looking future as scholars navigate this uncharted territory (Thumiki and Jurcic, 2021). It is time to recall the famous words of the American management guru: "Knowledge has to be improved, challenged, and raised constantly, or it vanishes." Knowledge management and its transfer have become critical in the current situation due to the swift disruption of markets and the economic concerns of individuals, families, and companies. The goal is to assess the effectiveness of the tools and techniques implemented and understand the challenges faced in the current pandemic. Knowledge management and transfer are most significant in any field of business, including services, because hackers are getting more creative and astute in creating threats to steal knowledge or data. Everyone faces severe challenges in protecting their understanding, and new security problems are constantly popping up.

In today's fast-changing world of business and technology, the importance of knowledge is becoming increasingly apparent. With the arrival of the information age, companies are reconsidering how they collect, save, and use knowledge to improve their work and develop new ideas (Mukanova et al., 2020). Knowledge management is a set of plans and ways to help a company effectively deal with information and expertise. To understand how knowledge management works today, it's essential to look back at how it developed over time and the situations it came from. The justification for undertaking a study on knowledge management is rooted in its pivotal role as a catalyst for organizational success. Nowadays, many industries rely a lot on knowledge, and companies that handle their knowledge well are better positioned to adapt, develop new ideas, and keep growing. This study wants to add to what the researchers already know by giving practical ideas about using knowledge management. The scholars also want to fill in the gaps in our understanding to see better how valuable good knowledge management is for all kinds of businesses. The researchers have taken up this research to give practical suggestions to companies that want to manage their knowledge better. This way, they can work more efficiently and be in an excellent position to compete in the ever-changing global business world.

13.2 LITERATURE REVIEW

One of the imposing "black swan" occurrences of the present century was the coronavirus disease (COVID-19) in December 2019, which turned into a global pandemic by March 2020 (Sohrabi et al., 2020). Due to the pandemic, the world's social, economic, and business life has reached a standstill. Quite a few countries enforced "lockdowns" to thwart people from being exposed to the virus (Lonska et al., 2021). As an effect, social distancing and virtual communication have become the order of the day. After a momentary pause, the premiers of all the nations switched their resources to fight the hidden opponent. The decrease in economic activity severely affected all the manufacturing and service sectors (retail, banks, educational institutions, real estate, IT, recreation, media, and healthcare) except for "front-line" workers who were fully engaged. The social distancing norms increased the use of the Internet and Internet-based services for communications and work-from-homes (D'Souza et al., 2023). In the aftermath of the COVID-19 pandemic, the knowledge management landscape has undergone a profound transformation. The pandemic has accelerated the need for effective knowledge transfer and management strategies, particularly in higher education institutions (HEIs) and corporate sectors. The disruptions caused by the pandemic underscore the importance of resilient knowledge systems that can adapt to unforeseen challenges. For specific periods, lectures worldwide have been conducted in classes, with teachers teaching in-person in their classrooms (Fong Boh et al., 2013). In the contemporary era, facilitated by advanced technology and internet accessibility, students can engage in education through online or digital formats. Even before the pandemic, online learning from courses on the Internet was complementing traditional, in-person lectures in various countries. Due to the protective procedures against the coronavirus, educational institutions in affected nations are transitioning to online formats to sustain teaching (Tomé et al., 2022). However, the effectiveness of this shift is significantly contingent on the availability and quality of technology, internet access, and internet speed. As indicated by Kemp (only approximately 60% of the global population is connected online. The transition to online education may be relatively seamless in developed nations with robust technological and Internet infrastructure.

The COVID-19 pandemic has harmfully disturbed educational institutions worldwide, resulting in repeated compulsory lockdowns and social distancing rules. After a short period, most organizations, including educational institutions, transferred their operations from established in-class to virtual, web-based teaching-learning. Due to exceptional change, faculty, students, and administrators faced several challenges. However, how educational institutions coped with this challenging situation remains an important question. To address this, the researchers from India present the Knowledge Management Process (KMP), Knowledge Management System Infrastructure (KMSINF), and Knowledge Management System Quality (KMSQU) as resilient strategies to convert challenges into opportunities. Using the KM processes and practices, a complex model is constructed to influence employee commitment, performance, and job satisfaction positively (Sivagnanam et al., 2023). The international epidemic has caused typical transformations in the functioning of several organizations, including HEI worldwide (Apte et al., 2022; Lonska et al., 2021). The employees, faculty (also called knowledge workers), students, and general public had to deal with health challenges imposed by the pandemic and, at the same time, continue to perform their regular activities (Henriksen et al., 2020). From now on, faculty will refer to "knowledge workers" who create, acquire, share, and disseminate knowledge among students. HEI had to understand ways to survive by formulating and implementing resilient strategies for HEI (Rana et al., 2020; Zieba & Bongiovanni, 2022). One of the substantial challenges educational institutions face during the crisis is to switch from in-class to web-based online education. In developing countries like India, such an immediate change was challenging because of the lack of adequate infrastructure to implement online instructions (Almaiah et al., 2020; Irshad et al., 2021; Usman et al., 2021).

On the contrary, adapting to distance teaching in less developed economies poses more significant challenges. HEIs have traditionally played a central role in knowledge transfer, actively engaging in synchronous knowledge dissemination. However, the pandemic has prompted a re-evaluation of traditional approaches (Wang and Wu,

2021). The shift toward remote learning and work has necessitated the development of innovative and adaptable knowledge transfer tools and techniques. As per the findings by Marie and Horváthová, each crisis has the potential to serve as a valuable wellspring of knowledge (Wang and Belardo, 2009). they concluded that noteworthy connections exist between knowledge management (KM) strategies and the characteristics inherent in a crisis. The importance of contextualized teaching and learning (CTL) becomes even more pronounced in the post-COVID-19 era. Rooted in real-world applications, CTL aligns with the evolving needs of learners who thrive in hands-on contexts. This approach is crucial in educational and corporate settings, where practical, applicable knowledge holds significant value.

The pandemic unleashed uncertainties around the world. Organizations scrambled to ensure business continuity. The pandemic was unprecedented as disruptions were far beyond business continuity scenarios planning concerning seriousness, timelines, and topographies. Primarily, business disruptions were assumed to last just a limited time at a regional controlling location. Conversely, the pandemic continued to extend, and efforts were needed to comply with ever-budding and modifying local, state, federal, and universal regulations, policies, and principles. Continued jobs at home befitted an obvious and merely way to protect business connections for many service-oriented organizations. Organizations not only tried to leverage existing KM practices but were also compelled to look at the efficacy of KM practices critically (Apte et al., 2023). The study was initiated in Iraq and investigated the relationship between KM processes and organizational sustainability in HEIs. It explores the moderating effect of online training on the relationship between KM and college sustainability. Documents and data for the research were amassed from respective HEIs in the Kurdistan area employing a survey questionnaire and a random sample technique. The findings confirm that knowledge codification/storage and knowledge generation substantially affect the region's sustainable development of HEIs. Further, staff training increases the understanding of KM processes, and knowledge utilization positively boosts the sustainability of institutions through the moderation of workshops. The scholars identified that information formation and expansion have a favorable consequence on additional knowledge management methods, inferring that HEIs ought to employ suitable tools to collect and generate information to improve the positive effects of other KM tools. The study's results hypothetically increase the knowledge of KM processes and their impacts on the sustainable development of universities in developing countries. (Budur et al., 2024).

Knowledge workers originate novel and potentially valuable concepts and resolutions for revitalizing a company's products, services, and processes (Amabile, 1988). Human Resource Management (HRM) assumes a crucial role in fortifying the organization's capacity for innovation by elevating the creativity of these knowledge workers. Furthermore, the collaboration between HEIs and businesses in the knowledge transfer process has gained heightened importance. The corporate sector's active involvement in training programs and workshops and continuous interaction with leading educational institutions exemplify a symbiotic relationship. This collaboration aims to foster mutual benefits, ensuring that knowledge dissemination remains relevant and responsive to the evolving demands of the post-pandemic world. The debate is part of a succession of webinars on school reopening in

Kazakhstan, structured in cooperation with the World Bank, UNICEF, UNESCO, WHO, and local partners (Bektursynova and Sarsengaliyeva, 2020). The goal is to empower education experts and professionals to enforce public health measures related to schools within the framework of COVID-19. The goal of empowering education includes minimizing the impact of learning disruptions and orchestrating digital and blended learning that prioritizes children and fosters inclusivity (Vuong et al., 2022). The dynamic stages of knowledge internalization and externalization remain integral components of the knowledge transfer process. As organizations strive to apply knowledge in daily operations, emphasizing work-based analysis and transforming tacit knowledge into explicit knowledge becomes paramount. The pandemic has highlighted the need for organizations to acquire knowledge and internalize and adapt it effectively to navigate the uncertainties of the post-COVID-19 landscape (Carreira, 2021). In the corporate sector, where knowledge is a critical asset, the pandemic has reinforced the importance of innovation and KM as core business strategies. Companies that proactively engage in these strategies are better positioned to create intellectual capital and enhance shareholder value, which is crucial for sustained success in the evolving business environment. The utilization of digital tools in managing knowledge activities holds the potential to improve organizational performance by fostering increased innovation capability. Implementing KM operates on the fundamental principle that individuals possessing knowledge are expected to share their insights with colleagues (Dul et al., 2011). This sharing aims to collectively enhance performance (Yakhlef, 2010) and establish a competitive advantage for the organization (Hislop et al., 2018). However, it is essential to note that the effectiveness of KM implementation across the organization is significantly influenced by individual differences. As the researchers navigate the complexities of a post-pandemic world, KM after COVID-19 should be viewed as a dynamic, adaptive process. Embracing innovative tools, fostering collaboration between HEIs and businesses, and recognizing the changing dynamics of knowledge absorption are vital elements in shaping effective KM strategies for the future. The lessons learned during the pandemic highlight the prominence of resilience, adaptability, and forward-thinking in managing and transferring knowledge in a world forever changed by the events of COVID-19 (Zhussupov et al., 2021). Potluri and Anjam (2021) accomplished similar kinds of research intending to explore different methods of knowledge transfer in the UAE HEIs (both pre- and post-COVID-19 pandemic) and to diagnose the indulgence level of faculty and students regarding these knowledge transfer methods.

13.3 CYBER SECURITY IN KNOWLEDGE MANAGEMENT

Security in KM has been imperative since the inception of the KM discipline, which has become a very effective way of managing a company's intellectual assets in present circumstances. From customer data to manuals and job aids, every company's KM system is an assortment of information that could offer a competitive advantage to any organization, which is an easy target for hackers. Concerning HEIs, policy documents, manuals, research data, accreditation process documents, evaluation criteria and their sheets, and other intellectual property should be protected more. It is a must for corporate and service sector companies to secure their KM system from physical and cyber threats. The higher education institution's KM system should consist of databases, content management, data warehouses, and wikis (user-friendly tools allowing anyone to upload, store, and access data in a central location). The quantity of knowledge an organization owns and how that knowledge is shared and employed have come to establish its success story, and KM principles were applied in ensuring cyber security compliance and whether such compliance inspired better or worse performance in the case study organization (Opoku-Ahene & Zwilling, 2022). The book chapter emphasizes the capability to comprehend and foresee your organization's part in a progressively multifaceted operating environment plays a crucial role in its sustained existence and claims that the usefulness of the critical asset of knowledge and the management of the vital resource plays an essential role in the accomplishment or catastrophe of this obligatory objective (Darby et al., 2021).

Cyber threats are gradually upsetting HEIs worldwide, and a surge in cyber-attacks has occurred in recent years, comprising data violations, ransomware outbreaks, phishing cheats, and malware defilements. The crucial driving factor for these attacks is the invaluable data these institutions hold, consisting of students', faculty's, and staff's sensitive personal information and, most significantly, research data and other intellectual property. HEIs with valuable data and information are the prime targets of cybercriminals. During and post-pandemic situations, which have given an experience of remote learning, further enhancing the risk of cyber-attacks on HEIs. Decentralized information technology (IT) infrastructure and the many devices connected to these university networks make these promising targets for cyber-attackers, making it challenging for them to keep their systems secure. Universities and colleges must protect their valuable data, intellectual property, and infrastructure from cyber-attacks. In HEIs, cyber-attacks are pretty acute due to the increase in the number and types of equipment that can be connected to data networks. The problem is influenced by educational and research activities involving many teaching staff, auxiliary or non-teaching staff and students, and complex computing and research systems in various fields (Cojocariu et al., 2020). The primary findings of another extensive literature review proved that empirical research on cyber security risks in higher education is scarce, and there are significant gaps in the literature. The paper also synthesizes an overview of mission-critical assets' everyday threat events, proposes a generic threat model, and summarizes common cybersecurity vulnerabilities (Ulven & Wangen, 2021).

13.4 POST-PANDEMIC KNOWLEDGE MANAGEMENT STRATEGIES IN HIGHER EDUCATION INSTITUTIONS (HEIs)

In the outburst of the pandemic in the entire world, the landscape of higher education in Kazakhstan underwent unprecedented challenges, prompting a re-evaluation of KM strategies within HEIs (Tastanbekova, 2019). Traditionally, these institutions have been viewed as vital contributors to societal development, fostering the growth of competent citizens capable of driving social welfare and productivity. The Kazakh government, aligning with this ethos, has meticulously promoted the advancement of education to cultivate a well-educated generation capable of navigating the rapidly evolving realms of scientific development and technological revolution. With a current count of 128 licensed scientific institutions (Kuzembayeva et al., 2022) offering 1,032 accredited education programs across various scientific and literary fields, Kazakhstan indicates its commitment to educational excellence. The Ministry of Higher Education and Scientific Research, in alignment with Sustainable Development Goal number four, seeks to confirm comprehensive, reasonable, and quality education while promoting lifelong learning opportunities (Kerimkulova and Kuzhabekova, 2017).

The existing body of literature exploring the interplay between KM and crisis management suggests that proficiently integrating KM practices is beneficial in navigating crises. In the study by Rosário et al. (2020), the authors delve into this relationship and offer insights into the advantageous effects of skillful KM implementation during periods of crisis. The response to the COVID-19 pandemic posed unique challenges to higher education. Yet, the Kazakh Ministry of Education displayed resilience by implementing quality-conscious approaches to uphold global standards across public, private, and international partnership institutions. Despite the pandemic's upheaval, the commitment to maintaining the quality of academic programs persisted, focusing on control mechanisms to ensure the satisfaction of all stakeholders involved, including students, graduates, families, and the broader Kazakh public. The underpinning theoretical framework for these initiatives manifests in the presented hypotheses, establishing connections between demographic factors and knowledge transfer (KT) variables within university settings (Haruna et al., 2022). These hypotheses form the foundation for comprehending and enhancing the KT process in the post-COVID-19 landscape:

1. The chosen field of study significantly shapes instructors' overall satisfaction with the KT process in Kazakhstani universities.
2. The educational attainment level of students plays a crucial role in determining their satisfaction levels with KT methods.
3. The field of study significantly influences students' overall satisfaction with the KT process in universities across Kazakhstan.
4. Choices in KT methods during the COVID-19 pandemic wield a substantial impact on the overall satisfaction levels of the student community.
5. The repercussions of employing online KT techniques significantly influence student satisfaction in Kazakhstan.
6. The diverse elements targeted for enhancement within the KT process significantly contribute to student satisfaction.
7. Challenges within the KT system significantly impact the overall satisfaction of the learning community.

As the higher education sector in Kazakhstan grapples with the complexities of the post-pandemic environment, considering these hypotheses is pivotal. (Durrani et al., 2023) It is essential to develop adaptive KM strategies that ensure the continual provision of high-quality education in alignment with global standards. The researchers have selected three hypotheses based on the comprehensive discussion above.

Hypotheses

H1: Time spent on self-study affects the success of passing the exam more than the variety of training formats.
H2: The quality of the Internet affects the quality of online education received in higher education institutions.
H3: Students' interaction and activity during traditional education significantly influence exam performance.

13.5 METHODOLOGY AND FINDINGS

The study involved 300 students from different faculties and years of study in various academic subjects. One hundred eighty-six students were boys, and 114 were girls, aged 16–23. The research was conducted based on Google Forms. In the first part of our study, a sample of people underwent multiple-choice questions where, for each question, people could choose one answer they considered the best among others. The questions were related to online education, offline education, or mixed education in different types of science, such as exact and inexact sciences. The effectiveness of other formats for teaching exact sciences, such as mathematics and physics, can vary depending on various factors, including the subject matter, the student's learning preferences, and the quality of instruction. It's essential to note that preferences and effectiveness may differ among individuals. Interestingly, social science and humanities have different results; depending on the subject, people's opinions vary. Furthermore, a sample of people explained their main reasons for choosing a specific answer, and the main common reasons for every answer were the same between all samples of people.

This sample of people, also consisting of students and professors, was divided into three main groups to continue ours. Group One (1) is online education, Group Two (2) is offline education, and Group Three (3) is mixed education. The sample of students for each of the three groups (online, offline, blended learning) was based on random selection from educational institutions, mainly in the largest city of Kazakhstan, Almaty. The researchers considered the various specializations of universities in which multiple types of lessons and levels of education are held from the first to the fourth year of study to achieve representativeness of the sample as shown in Table 13.1. The exams were given based on relevant background and covered the key topics of each course.

Exams are governed in monitored conditions to ensure that results are protected. Based on the results of the exams, we conducted a statistical analysis of additional factors, such as the time students spent studying subjects. All students gave their answers based on their average time studying different subjects. The results of exams show the measure of our success. The data were carefully coded and decoded, and Microsoft Excel and SPSS software were applied to identify patterns and draw conclusions. The data of online education, traditional education, and mixed education was taken as a Dummy variable (variable can be zero or one) where online education is 0, and mixed education is 1. Statistical analysis methods were used to identify statistically significant differences between groups, including means, standard deviations, and correlation analysis as shown in Table 13.2.

TABLE 13.1
Regression Statistics

Multiple R	0.94530907
R Square	0.89360923
Adjusted r square	0.8836351
Standard Error	2.09937296
Observation	300

TABLE 13.2
Regression Analysis (N = 300)

	df	SS	MS	F	Significance F
Regression	3	1184.60315	394,867717	89,5926595	1,1804E-15
Residual	32	141,035739	4,40736684		
Total	35	1325,63889			

Source: Research findings.

The collected data showed research results with the R square coefficient, the significance of F and P- values of three different education types—online, mixed, and traditional. Our significance 1.1804E-15 means that the provided summary output appears to be from a regression analysis, likely conducted using statistical software such as R or Python, focusing on data science. Let's break down the key elements and interpret them in context:

Regression analysis

- Multiple *R*: This value (0.945) represents the correlation between the dependent and independent variables. A high value indicates a strong linear relationship.
- *R* Square (R^2): (0.894) signifies the relationship of the variation in the dependent variable that is predictable from the independent variables.
- Standard Error: The standard deviation of the residuals is 2.099, indicating the average distance between observed and predicted values.

ANOVA (Analysis of Variance)

- The table breaks down the variance into regression and residual components.
- The F-statistic (89.59) with a very low p-value (1.18E-15) suggests that at least one independent variable significantly contributes to the model.

Coefficients

- Intercept: The starting point of the regression line is 43.74. The associated statistics provide information about its significance and confidence interval.
- (Online), (Mix), (Time): These are the coefficients for the respective independent variables. They show the impact on the dependent variable. Associated statistics include standard error, t-statistic, and p-value.

In summary, the regression model seems to have high explanatory power (R Square) and significant predictors based on the low p-values in the ANOVA table. The coefficients for each variable provide insights into their contributions. The model includes three independent variables (Online, Mix, and Time) and an intercept.

As shown in Table 13.1, as R Square explains, our forecasting of 89% means that the table is helpful. Based on the above data, the researchers can assume that the quality of KT does not directly depend on different learning formats, such as online, offline, or a mix of learning. Based on our research, scholars can assume that the time spent studying affects education results more, but the form of education helps improve the information received. Online education shows the best result with a coefficient of 0.917, meaning that online education during COVID-19 has developed significantly and brings a more productive methodology for transferring knowledge to HEIs from Almaty, Kazakhstan. Furthermore, mixed education bears fruit and shows that HEIs took only the best processes from the two learning formats after the pandemic. The third part of our study included online learning participants of different levels of training. They assessed the quality of the Internet used during online classes.

13.6 CONCLUSION

In conclusion, our exploration of KM in higher institutions, conducted through a comprehensive blend of online, offline, and mixed-mode education, has shed light on critical information handling and dissemination aspects. The findings underscore the growing importance of efficient KM systems in academic settings. In carrying out this study, it is recommended that educational institutions pay more attention to factors that can affect learning, such as the Internet, the professionalism of teachers, and other factors that can influence learning. On the part of students, it should be noted that the time spent on studying significantly affects the success of exams and understanding of information. Even with this information, the researchers must recognize the study's limitations. The reliance on a specific collection approach, whether online, offline, or mixed, may introduce biases or limitations in capturing the full spectrum of KM practices. The sample size and geographical scope of our study may also affect the generalizability of the findings. New research should address these limitations by employing diverse data collection methods, expanding the participant pool, and considering a broader range of institutional contexts.

Most of the academic institutions in Kazakhstan are followed by the expertise offered by the Cisco Networking Academy, Huawei, and Kaspersky Lab to protect their soft versions of data from cyber threats. Students actively participate in scientific and innovative activities related to this discipline. The entire education sector of the country, keeping in mind the Ministry of Digital Development, Innovation, and

Aerospace Industry of the Republic of Kazakhstan approved the "Cyber Shield of Kazakhstan" cybersecurity concept and developed various cybersecurity programs to protect their information technology and computing sections. As of today, Kazakhstan has about 40 companies, 19 private security operational centers (SOCs), 3 computer incident response teams (CERTs), 7 private, accredited testing laboratories, 8 higher educational institutions, and 25 secondary educational institutions, dealing with cybersecurity matters. The Ministry identified problems related to insufficient awareness among citizens, shortage of information security professionals, inadequate infrastructure, limited trust in the public sector, and risks associated with the provision of electronic public services. Based on the identification of the above risks, the country is persistently reviewing the current cybersecurity situation, identifying key challenges and threats, learning from the experiences of other countries, framing objectives, and formulating an action plan (International Telecommunicaiton Union, 2022).

Regarding academic mobility, students and master's students study at foreign universities in the USA, Europe (Ukraine, Poland, Italy, France, Germany, Portugal), CIS, and Asia (Korea, Malaysia, China). Currently, the department prepares specialists in two directions: information security and information systems at all levels of education. The educational program for information security includes the following directions: Security of Systems and Networks, Cryptographic Information Protection, Technical Information Protection, and Blockchain Engineering. The educational program for information systems consists of the following trajectories: DevOps, Business Analytics and IT Project Management, IS Architecture, and Business Application Development (Satbayev University, 2024). The study on KM in higher institutions opens avenues for future research with the potential to deepen our understanding and refine existing practices. Researchers are encouraged to delve into specific aspects, such as internal factors that can be changed, and always look at the problem from both sides, both from professors and students.

REFERENCES

Alin-Ciprian Cojocariu, Ion Verzea, & Rachid Chaib (2020). Aspects of cyber-security in higher education institutions. In *Proceedings of International Symposium in Management (SIM2019) Innovation for Sustainable Management and Entrepreneurship* (pp. 3–11). Springer Proceedings in Business and Economics Book Series (SPBE).

Almaiah, M. A., Al-Khasawneh, A., & Althunibat, A. (2020). Exploring the critical challenges and factors influencing the E-learning system usage during the COVID-19 pandemic. *Education and Information Technologies*, *25*, 5261–5280.

Amabile, T. M. (1988). A model of creativity and innovation in organizations. *Research in Organizational Behavior*, *10*(1), 123–167.

Apte, S., Lele, A., & Choudhari, A. (2022). COVID-19 pandemic influence on organizational knowledge management systems and practices: Insights from an Indian engineering services organization. *Knowledge and Process Management*, 1–13, https://doi.org/10.1002/kpm.1711

Apte, S., Lele, A., & Choudhari, A. (2023). COVID-19 pandemic influence on organizational knowledge management systems and practices: Insights from an Indian engineering services organization. *Knowledge and Process Management*, *30*(3), 254–266.

Arcia, G., De Hoyos, R., Patrinos, H., Sava, A., Shmis, T., & Teixeira, J. (2021). *Learning recovery after COVID-19 in Europe and* Policy and Practice.

Bektursynova, A., & Sarsengaliyeva, B. (2020, June). Impact of terrible pandemic COVID-19 on Kazakhstani education and it's outcome. In *Proceedings of International Young Scholars Workshop* (Vol. 9).

Budur, T., Abdullah, H., Rashid, C. A., & Demirer, H. (2024). The connection between knowledge management processes and sustainability at higher education institutions. *Journal of the Knowledge Economy*, *51*, 1–34.

Carreira, M. I. P. G. (2021). *A model for enterprise knowledge management after covid-19 crisis* (Doctoral dissertation, Universidade NOVA de Lisboa (Portugal)).

Darby, R., Dodd, L., & Hilton, J. (2021). Cyber security and knowledge management. In *Managing security* (pp.122–142). Routledge.

D'Souza, G. S., Irudayasamy, F. G., & Parayitam, S. (2023). Emotional exhaustion, emotional intelligence and task performance of employees in educational institutions during COVID-19 global pandemic: A moderated-mediation model. *Personnel Review*, *52*(3), 539–572.

Dul, J., Ceylan, C., & Jaspers, F. (2011). Knowledge workers' creativity and the role of the physical work environment. *Human Resource Management*, *50*(6), 715–734.

Durrani, N., Qanay, G., Mir, G., Helmer, J., Polat, F., Karimova, N., & Temirbekova, A. (2023). Achieving SDG 4, Equitable quality education after COVID-19: Global evidence and a case study of Kazakhstan. *Sustainability*, *15*(20), 14725.

Fong Boh, W., Nguyen, T. T., & Xu, Y. (2013). Knowledge transfer across dissimilar cultures. *Journal of Knowledge Management*, *17*(1), 29–46.

Haruna, U. A., Amos, O. A., Gyeltshen, D., Colet, P., Almazan, J., Ahmadi, A., & Sarria-Santamera, A. (2022). Towards a post-COVID world: Challenges and progress of recovery in Kazakhstan. *Public Health Challenges*, *1*(3), e17.

Henriksen, D., Creely, E., & Henderson, M. (2020). Folk pedagogies for teacher transitions: Approaches to synchronous online learning in the wake of COVID-19. *Journal of Technology and Teacher Education*, *28*(2), 201–209.

Hislop, D., Bosua, R., & Helms, R. (2018). *Knowledge management in organizations: A critical introduction*. Oxford University Press.

International Telecommunicaiton Union (2022). s. https://www.itu.int/hub/2022/08/implementing-kazakhstan-cybersecurity/

Irshad, H., Umar, K. M., Rehmani, M., Khokhar, M. N., Anwar, N., Qaiser, A., & Naveed, R. T. (2021). Impact of work-from-home human resource practices on the performance of online teaching faculty during coronavirus disease 2019. *Frontiers in Psychology*, **12**, 740644.

Kerimkulova, S., & Kuzhabekova, A. (2017). Quality assurance in higher education of Kazakhstan: a review of the system and issues. *The Rise of Quality Assurance in Asian Higher Education*, *11*, 87–108.

Kuzembayeva, G., Umarova, A., Maydangalieva, Z., Gorbatenko, O., Kalashnikova, E., Kalmazova, N., & Chigisheva, O. (2022). Content and language integrated learning practices in Kazakhstan secondary schools during the COVID-19 pandemic. *Contemporary Educational Technology*, *14*(2), ep362.

Lonska, J., Mietule, I., Litavniece, L., Arbidane, I., Vanadzins, I., Matisane, L., & Paegle, L. (2021). Work–life balance of the employed population during the emergency situation of COVID-19 in Latvia. *Frontiers in Psychology*, *12*, 682459.

Mukanova, P., Eftimova, S., & Mincheva, K. (2020). Facebook and Bulgarian libraries: The challenges of modern information environment. In *EDULEARN20 Proceedings* (pp. 890–898). IATED.

Opoku-Ahene, A. R., & Zwilling, M. (2022). The influence of cyber security implementation strategy on organizational knowledge management and performance–A case study of Sinapi Aba savings and loans in Ghana. *Digital Transformation: The Harmonic Convergence of People, Culture, Process, and Technology in the New Normal*, To Know Press.

Potluri, R. M., & Anjam, M. (2021, January). Knowledge transfer in UAE higher education institutions during Covid-19 pandemic: Learners' cannot learn surgery by watching. In *2021, the 2nd International Conference on Computation, Automation and Knowledge Management (ICCAKM)* (pp. 25–30). IEEE.

Rana, S., Anand, A., Prashar, S., & Haque, M. M. (2020). A perspective on the positioning of Indian business schools post-COVID-19 pandemic. *International Journal of Emerging Markets*, *17*(2), 353–367.

Rosário, A. M. F. T., Vilaça, F. I. D. A. F., Raimundo, R., & Cruz, R. M. N. (2020). Literature review on health knowledge management in the last 10 years (2009-2019). *Electronic Journal of Knowledge Management*, *18*(3), 338–355.

Satbayev University (2024). Department of cybersecurity, information processing and storage. https://official.satbayev.university/en/information-telecommunication-technologies/isec

Sivagnanam, P., Pillai, A. R., Elangovan, R., & Parayitam, S. (2023). Knowledge management process, infrastructure, and system quality as resilient strategies to respond to COVID-19 pandemic challenges: Evidence from higher educational institutions in India. *Knowledge and Process Management*, *30*(4), 333–354.

Sohrabi, C., Alsafi, Z., O'Neill, N., Khan, M., Kerwan, A., Al-Jabir, A., …, & Agha, R. (2020). World Health Organization declares global emergency: A review of the 2019 novel coronavirus (COVID-19). *International Journal of Surgery*, *76*, 71–76.

Tastanbekova, K. (2019). Student engagement in quality assurance of higher education in Kazakhstan. *Student Engagement and Quality Assurance in Higher Education: International Collaborations for the Enhancement of Learning* Routledge, (p. 109).

Thumiki, V. R. R., & Jurcic, A. (2021). Impact of COVID-19 crisis on knowledge management practices in Sultanate of Oman. *Electronic Journal of Knowledge Management*, *19*(3), 213–225.

Tomé, E., Gromova, E., & Hatch, A. (2022). Knowledge management and COVID-19: Technology, people and processes. *Knowledge and Process Management*, *29*(1), 70–78.

Ulven, J. B., & Wangen, G. (2021). A systematic review of cybersecurity risks in higher education. *Future Internet*, *13*(2), 39.

Usman, S. A., Kowalski, K. B., Andiappan, V. S., & Parayitam, S. (2021). Effect of knowledge sharing and interpersonal trust on psychological capital and emotional intelligence in higher-educational institutions in India: Gender as a moderator. *FIIB Business Review 19*, 366–381.

Vuong, Q. H., Le, T. T., La, V. P., Nguyen, H. T. T., Ho, M. T., Van Khuc, Q., & Nguyen, M. H. (2022). Covid-19 vaccines production and societal immunization under the serendipity-mindsponge-3D knowledge management theory and conceptual framework. *Humanities and Social Sciences Communications*, *9*(1), 181–189.

Wang, W. T., & Wu, S. Y. (2021). Knowledge management based on information technology in response to COVID-19 crisis. *Knowledge Management Research & Practice*, *19*(4), 468–474.

Wang, W. T., & Belardo, S. (2009). The role of knowledge management in achieving effective crisis management: A case study. *Journal of Information Science*, *35*(6), 635–659.

Yakhlef, A. (2010). The corporeality of practice-based learning. *Organization Studies*, *31*(4), 409–430.

Zieba, M., & Bongiovanni, I. (2022). Knowledge management and knowledge security—Building an integrated framework in the light of COVID-19. *Knowledge and Process Management*, *1–11*, 121–131.

Zhussupov, B., Saliev, T., Sarybayeva, G., Altynbekov, K., Tanabayeva, S., Altynbekov, S., & Fakhradiyev, I. (2021). Analysis of COVID-19 pandemics in Kazakhstan. *Journal of Research in Health Sciences*, *21*(2), e00512.

14 Future Directions and Emerging Trends in Cybersecurity for Knowledge Management

Philip Eappen, Adeyemi Abel Ajibesin, and Narasimha Rao Vajjhala

14.1 INTRODUCTION

The current state of cybersecurity in knowledge management (KM) is a critical area of focus as organizations increasingly rely on digital systems to manage and share knowledge. The integration of information and communication technology (ICT) into KM has brought significant advantages, including improved efficiency and accessibility of information. However, it has also introduced various vulnerabilities and risks associated with the security of organizational knowledge. As organizations increasingly adopt digital KM systems, the potential for cyber threats has escalated, necessitating a robust cybersecurity framework. According to Wang and Wang (2021), KM can play a pivotal role in addressing cybersecurity issues by emphasizing the interrelated social, organizational, and technological factors involved in cybersecurity. This sociotechnical systems approach highlights the need for a comprehensive understanding of how KM systems can be leveraged to enhance cybersecurity resilience (Wang & Wang, 2021).

KM systems are susceptible to various security issues, including unauthorized access, data breaches, and knowledge theft. The rapid digitization of knowledge assets has made them attractive targets for cybercriminals. For instance, Şandor and Tonţ (2021) emphasize that the vulnerabilities introduced by ICT can lead to significant risks regarding the security of information and organizational knowledge capital, which can result in information theft and loss of competitive advantage. The complexity of KM systems often leads to an uneven distribution of vulnerabilities, as highlighted by Wang et al. (2023), who discusses the challenges of managing vulnerabilities in interconnected systems. This complexity necessitates a proactive approach to identify and mitigate potential vulnerabilities in KM systems. Various approaches and solutions have been proposed to enhance cybersecurity in KM. One significant approach is the development of comprehensive risk management frameworks that integrate cybersecurity considerations into KM practices. For example, the National Institute of Standards and Technology (NIST) has developed a Cybersecurity

DOI: 10.1201/9781003498094-14

Framework that provides guidelines for managing cybersecurity risks, which can be adapted for KM contexts (Zimmerman & Zimmerman, 2019). Additionally, the implementation of vulnerability assessment tools and techniques is crucial for identifying and addressing potential security weaknesses. Walkowski et al. (2020) emphasize the importance of real-time vulnerability assessments to reduce the likelihood of successful cyberattacks.

Moreover, organizations are increasingly recognizing the need for cybersecurity awareness and training programs to equip employees with the knowledge and skills necessary to protect sensitive information. Garba et al. (2020) conducted a study on cybersecurity awareness among university students, highlighting the importance of education in mitigating cyber threats. The integration of artificial intelligence and machine learning into KM systems can enhance security by enabling predictive analytics and automated threat detection, as discussed by Althar et al. (2023). As organizations continue to leverage digital KM systems, it is imperative to adopt comprehensive cybersecurity strategies that encompass risk assessment, employee training, and the integration of advanced technologies. By addressing these challenges, organizations can enhance their cybersecurity posture and protect their valuable knowledge assets.

14.2 EMERGING TECHNOLOGIES IN CYBERSECURITY FOR KNOWLEDGE MANAGEMENT

The rapid evolution of technology has significantly transformed the landscape of cybersecurity, particularly in KM. As organizations increasingly rely on digital platforms to store, share, and manage knowledge, the need for robust cybersecurity measures has become paramount. This chapter explores three emerging technologies in cybersecurity—Artificial Intelligence and Machine Learning (AI/ML), Blockchain and Distributed Ledger Technology, and Quantum Computing and Quantum-Resistant Encryption—and their implications for enhancing security in KM systems.

14.2.1 Artificial Intelligence and Machine Learning (AI/ML)

AI and ML have emerged as powerful tools in the cybersecurity arsenal, particularly for KM systems. These technologies enable organizations to detect anomalies, model potential threats, and automate responses to cyber incidents. AI/ML algorithms can analyze vast amounts of data in real time, identifying patterns and anomalies that may indicate a security breach. For instance, Rizvi (2023) highlights how AI-based systems are increasingly employed for threat detection and prevention, allowing organizations to respond swiftly to potential threats. Predictive threat modeling using AI/ML can enhance an organization's ability to foresee and mitigate risks before they materialize. By analyzing historical data and identifying trends, these technologies can provide insights into potential vulnerabilities and attack vectors. Sarker et al. (2021) emphasize that AI-driven cybersecurity solutions can significantly improve the detection and response capabilities of organizations, particularly in complex environments where traditional methods may fall short.

The integration of AI/ML into cybersecurity frameworks offers several benefits for KM systems. First, these technologies enable organizations to enhance their situational awareness by providing real-time insights into potential threats. This proactive approach allows organizations to allocate resources more effectively and prioritize their cybersecurity efforts. Additionally, AI-driven insights can facilitate the identification of human errors, which are often the root cause of security breaches. AI/ML can automate routine cybersecurity tasks, freeing up human resources to focus on more strategic initiatives. This automation not only improves efficiency but also reduces the likelihood of human error, which is a significant factor in many cybersecurity incidents. As organizations continue to grapple with the increasing complexity of cyber threats, the adoption of AI-driven solutions will be essential for maintaining robust cybersecurity in KM systems.

14.2.2 Blockchain and Distributed Ledger Technology

Blockchain technology has gained traction as a secure and decentralized method for managing knowledge and data integrity. By providing a tamper-proof ledger of transactions, blockchain ensures that knowledge assets are securely shared and maintained. This decentralized approach mitigates the risks associated with centralized data storage, such as data breaches and unauthorized access. Adegbite et al. (2023) discuss how blockchain can enhance cybersecurity by providing a transparent and immutable record of data transactions, thereby increasing trust among stakeholders. In KM, blockchain can facilitate the secure sharing of sensitive information among authorized users while maintaining data integrity. For example, organizations can use blockchain to track the provenance of knowledge assets, ensuring that only verified and authorized information is shared. This capability is particularly valuable in industries where data integrity is critical, such as healthcare and finance.

While blockchain offers significant advantages for KM security, it is not without limitations. One potential application of blockchain in KM is the creation of decentralized knowledge repositories, where users can contribute and access information without relying on a central authority. However, the implementation of blockchain technology can be complex and resource-intensive, requiring significant investment in infrastructure and training. The scalability of blockchain solutions remains a concern, particularly in environments with high transaction volumes. As noted by Pooyandeh et al. (2022), the performance of blockchain networks can be hindered by their consensus mechanisms, which may limit their applicability in real-time KM scenarios. Additionally, the regulatory landscape surrounding blockchain technology is still evolving, which may pose challenges for organizations seeking to implement these solutions.

14.2.3 Quantum Computing and Quantum-Resistant Encryption

Quantum computing represents a paradigm shift in computational capabilities, with the potential to disrupt traditional encryption methods. Quantum computers can solve complex mathematical problems at unprecedented speeds, rendering many current encryption algorithms vulnerable to attacks. As quantum computing technology advances, organizations must prepare for the implications it poses for cybersecurity in KM. The ability of quantum computers to break widely used encryption standards, such as RSA and ECC, necessitates the development of quantum-resistant encryption methods. Malik (2021) emphasizes the urgency of addressing the risks posed by quantum computing to ensure the continued security of sensitive information. Organizations must begin transitioning to quantum-resistant algorithms to safeguard their knowledge assets against future threats.

In response to the challenges posed by quantum computing, researchers and organizations are actively developing quantum-resistant encryption algorithms. These algorithms are designed to withstand the computational power of quantum computers, ensuring the security of sensitive data in the post-quantum era. Taddeo et al. (2019) discuss the importance of investing in quantum-resistant encryption technologies to future-proof cybersecurity measures. The development of quantum-resistant encryption is critical for KM systems, where the confidentiality and integrity of information are paramount. By adopting these advanced encryption techniques, organizations can mitigate the risks associated with quantum computing and ensure the continued protection of their knowledge assets.

The integration of emerging technologies such as AI/ML, blockchain, and quantum computing into cybersecurity frameworks presents significant opportunities for enhancing the security of KM systems. AI/ML technologies enable organizations to detect threats, model risks, and automate responses, while blockchain offers a secure and decentralized approach to knowledge sharing. Additionally, the development of quantum-resistant encryption is essential for safeguarding sensitive information in the face of advancing quantum computing capabilities. As organizations continue to navigate the complexities of the digital landscape, the adoption of these emerging technologies will be crucial for maintaining robust cybersecurity in KM.

14.3 ADVANCED CYBERSECURITY APPROACHES FOR KNOWLEDGE MANAGEMENT

In the contemporary digital landscape, the protection of KMS is paramount, as these systems often house sensitive organizational data. Advanced cybersecurity approaches, such as Zero Trust Architecture (ZTA), Privacy-Enhancing Technologies (PETs), and Cybersecurity Automation and Orchestration, are essential for safeguarding these systems against unauthorized access and internal threats. ZTA represents a paradigm shift in cybersecurity, emphasizing that no entity, whether inside or outside the network, should be trusted by default. This approach is particularly relevant in KM, where sensitive information is frequently accessed by various stakeholders. The core principles of ZTA include continuous verification, least privilege access, and micro-segmentation of networks (Cao et al., 2024). By implementing these principles, organizations can significantly reduce the risk of data breaches and ensure that only authorized personnel can access sensitive information. The application of ZTA in KM systems enhances security by creating a robust framework that continuously monitors user behavior and access patterns. This proactive stance allows organizations to detect anomalies in real time, thereby preventing unauthorized access and mitigating potential internal threats (Pokhrel et al., 2023). For instance, by employing identity and access management tools within a ZTA framework, organizations can enforce strict access controls based on user roles and responsibilities, ensuring that sensitive knowledge is only accessible to those who require it for their work (Cao et al., 2024). The advantages of ZTA are manifold. By eliminating the traditional perimeter-based security model, ZTA reduces the attack surface and enhances the organization's ability to respond to threats. Furthermore, it fosters a culture of security awareness among employees, as they are required to adhere to strict access protocols (Pulyala et al., 2019). This shift not only protects sensitive knowledge but also promotes a more secure organizational environment overall.

Privacy-Enhancing Technologies (PETs) play a crucial role in balancing data privacy with secure collaborative KM. Technologies such as homomorphic encryption, federated learning, and secure multi-party computation enable organizations to process and analyze data without exposing sensitive information (Zhang et al., 2021). For example, homomorphic encryption allows computations to be performed on encrypted data, ensuring that sensitive information remains confidential even during processing (Hosseini et al., 2023). Federated learning, another significant PET, facilitates decentralized data processing, allowing organizations to collaboratively train machine learning models without sharing raw data. This approach is particularly beneficial for KM systems, as it enables organizations to leverage collective insights while maintaining data privacy (Zhang et al., 2021). By employing federated learning, organizations can enhance their KM capabilities while ensuring compliance with data protection regulations. The integration of PETs into KM systems not only enhances data privacy but also fosters trust among stakeholders. By demonstrating a commitment to protecting sensitive information, organizations can encourage collaboration and knowledge sharing, ultimately leading to more innovative solutions and improved decision-making (Zhang et al., 2021). However, the implementation of these technologies requires careful consideration of the associated challenges, such as computational overhead and the need for robust security measures to protect the underlying infrastructure (Hosseini et al., 2023).

Cybersecurity automation and orchestration are critical components in managing cybersecurity threats in real time, especially within large-scale KM systems. Automated response systems can significantly enhance an organization's ability to detect and respond to threats swiftly, thereby minimizing potential damage (Norem et al., 2021). By automating routine security tasks, organizations can free up valuable resources and allow cybersecurity professionals to focus on more complex issues. The role of Security Orchestration, Automation, and Response (SOAR) tools is particularly noteworthy in this context. SOAR platforms integrate various security tools and processes, enabling organizations to respond to incidents more effectively (Pulyala et al., 2019). For instance, by automating the incident response process, organizations can reduce response times and improve their overall security posture. This is especially crucial in KM systems, where timely access to information is vital for decision-making. However, the implementation of cybersecurity automation is not without its challenges. Organizations must navigate issues such as the potential for false positives, the complexity of integrating disparate systems, and the need for continuous monitoring and updates to automated processes (Norem et al., 2021). Moreover, while automation can enhance efficiency, it is essential to maintain human oversight to ensure that automated systems are functioning as intended and to address any unforeseen issues that may arise (Pulyala et al., 2019).

The benefits of cybersecurity automation and orchestration in KM systems are substantial. By enhancing the speed and efficiency of threat detection and response, organizations can better protect sensitive knowledge and maintain the integrity of their information systems (Norem et al., 2021). The ability to automate routine tasks allows cybersecurity teams to allocate their resources more effectively, ultimately leading to a more resilient organizational framework. Advanced cybersecurity approaches such as ZTA, PETs, and Cybersecurity Automation and Orchestration are essential for safeguarding KM systems. By implementing these strategies, organizations can enhance their security posture, protect sensitive information, and foster a culture of collaboration and trust. As the digital landscape continues to evolve, the adoption of these advanced cybersecurity measures will be critical in ensuring the integrity and confidentiality of organizational knowledge.

14.4 ETHICAL AND REGULATORY CONSIDERATIONS IN CYBERSECURITY FOR KNOWLEDGE MANAGEMENT

In the rapidly evolving landscape of cybersecurity, the integration of ethical and regulatory considerations is paramount, especially in KM. As organizations increasingly rely on digital platforms to store and manage sensitive information, the implications of data privacy regulations, ethical AI practices, and transparency in cybersecurity practices become critical. Data privacy regulations such as the General Data Protection Regulation (GDPR) and the California Consumer Privacy Act (CCPA) have significantly impacted KM practices. The GDPR, which came into effect in May 2018, mandates strict guidelines for the collection, processing, and storage of personal data. It emphasizes the principles of data minimization, purpose limitation, and user consent, thereby requiring organizations to implement robust data protection measures. Similarly, the CCPA provides California residents with rights regarding their personal data, including the right to know what data is collected and the right to delete it. These regulations compel organizations to adopt comprehensive data governance strategies, which are essential for ensuring compliance and protecting sensitive information within KM systems.

Best practices for compliance with these regulations include conducting regular data audits, implementing strong encryption protocols, and ensuring that data processing activities are transparent to users (Determann, 2019). Organizations should also establish clear data retention policies and provide training to employees on data protection practices. By integrating these measures into emerging cybersecurity frameworks, organizations can not only comply with legal requirements but also enhance their overall security posture. Furthermore, organizations must remain vigilant about evolving regulations and adapt their practices accordingly to mitigate the risk of noncompliance and potential penalties. The increasing reliance on AI in cybersecurity raises significant ethical considerations. AI-driven tools are often employed for threat detection and response, but they can inadvertently perpetuate biases present in the training data. This bias can lead to discriminatory practices in decision-making processes, particularly in identifying potential threats (Rangaraju, 2023). Addressing these ethical concerns requires organizations to prioritize transparency and fairness in the development and deployment of AI systems. For instance, employing diverse datasets and implementing bias mitigation strategies can help ensure that AI tools operate equitably across different demographic groups (Rangaraju, 2023).

Moreover, ethical considerations in AI-powered cybersecurity extend to the decision-making processes within these systems. Organizations must establish ethical guidelines that govern the use of AI, ensuring that decisions made by automated systems align with broader ethical principles (Komaruddin et al., 2023). This includes considerations around privacy, accountability, and the potential consequences of automated decisions. By fostering a culture of ethical awareness and responsibility, organizations can enhance the integrity of their cybersecurity practices and build trust among stakeholders (Rangaraju, 2023). Building trust in cybersecurity practices is essential for effective KM, and transparency plays a crucial role in this process. Organizations must communicate their cybersecurity policies and practices clearly to stakeholders, including employees, customers, and partners. This transparency not only fosters trust but also encourages collaboration in maintaining security. For example, organizations can provide regular updates on cybersecurity incidents and the measures taken to address them, thereby demonstrating accountability and commitment to data protection.

Accountability in cybersecurity is equally important, as it ensures that organizations take responsibility for their actions and decisions related to data protection. This includes establishing clear roles and responsibilities for cybersecurity governance and ensuring that there are mechanisms in place for reporting and addressing security breaches (Morrow & Fitzpatrick, 2020). By promoting a culture of accountability, organizations can enhance their resilience against cyber threats and ensure that their KM systems remain secure. The integration of ethical considerations into cybersecurity practices can enhance accountability. Organizations should adopt a framework for cybersecurity ethics, which emphasizes the importance of beneficence, autonomy, and justice in decision-making processes (Formosa et al., 2021). By aligning their cybersecurity practices with these ethical principles, organizations can ensure that they act responsibly and uphold the rights and interests of individuals whose data they manage.

The ethical and regulatory considerations in cybersecurity for KM are multifaceted and critical for ensuring the protection of sensitive information. Compliance with data privacy regulations such as GDPR and CCPA is essential for safeguarding personal data and maintaining trust among stakeholders. Additionally, addressing ethical concerns in AI-powered cybersecurity tools and promoting transparency and accountability are vital for fostering a secure and trustworthy environment for KM. As organizations navigate the complexities of the digital landscape, the integration of these considerations will be crucial for enhancing their cybersecurity practices and ensuring the integrity of their KM systems.

14.5 CHALLENGES AND LIMITATIONS OF EMERGING CYBERSECURITY TRENDS

As organizations increasingly adopt emerging cybersecurity technologies to protect their KM systems, they face a myriad of challenges and limitations. These challenges encompass integration and scalability issues, cost and resource implications, and the complexity of new cybersecurity measures. Addressing these challenges is crucial for organizations to effectively safeguard sensitive information while ensuring operational efficiency. One of the primary challenges organizations encounter when implementing emerging cybersecurity technologies is the integration of these new systems with existing infrastructure. Many organizations operate on legacy systems that may not be compatible with modern cybersecurity solutions, leading to significant integration hurdles. For instance, integrating advanced technologies such as AI and ML into existing security frameworks can be complex and resource-intensive. Legacy systems may lack the necessary APIs or data formats required for seamless integration, resulting in fragmented security measures that can leave vulnerabilities unaddressed.

Additionally, scalability poses a significant concern for organizations as they expand their cybersecurity measures across diverse environments. Organizations often operate in heterogeneous environments, encompassing various platforms, devices, and applications. This diversity complicates the deployment of uniform cybersecurity measures, as different systems may require tailored solutions to address specific vulnerabilities (Coventry & Branley, 2018). For example, the healthcare sector, which often utilizes a mix of legacy and modern systems, faces unique challenges in scaling cybersecurity measures effectively (Coventry & Branley, 2018). As organizations grow and evolve, ensuring that cybersecurity solutions can scale accordingly without compromising security becomes increasingly critical.

The financial, technical, and human resources required for implementing emerging cybersecurity technologies can be substantial. Organizations must invest in not only the technology itself but also in the necessary infrastructure, training, and ongoing maintenance. For instance, deploying AI-driven cybersecurity solutions often requires significant investments in hardware and software, as well as skilled personnel to manage and operate these systems (Pulyala et al., 2019). The high costs associated with acquiring and maintaining advanced cybersecurity technologies can strain budgets, particularly for smaller organizations that may lack the financial resources to support such initiatives. Moreover, organizations must balance their cybersecurity investments with expected outcomes in KM. The challenge lies in justifying the return on investment (ROI) for cybersecurity expenditures, especially when the benefits of enhanced security are often intangible and difficult to quantify (Kour et al., 2019). Organizations may struggle to demonstrate the value of their cybersecurity investments to stakeholders, leading to potential underfunding of critical security initiatives. This financial pressure can result in organizations prioritizing short-term cost savings over long-term security investments, ultimately exposing them to greater risks.

The complexity of new cybersecurity measures can hinder their usability and effectiveness. Emerging technologies often come with intricate configurations and operational requirements that can overwhelm users and administrators alike. For instance, advanced threat detection systems powered by AI may generate a high volume of alerts, making it challenging for security teams to prioritize and respond to genuine threats effectively (Stafford et al., 2018). The complexity of these systems can lead to alert fatigue, where security personnel become desensitized to notifications, potentially resulting in missed critical alerts. User adaptation is another significant challenge in the successful implementation of emerging cybersecurity technologies. Organizations must invest in comprehensive training programs to ensure that employees understand how to use new systems effectively and adhere to security protocols. Without adequate training, users may inadvertently compromise security by failing to follow best practices or misconfiguring systems (Arafa et al., 2023). The human factor remains a critical aspect of cybersecurity, as employees are often the first line of defense against cyber threats. Therefore, fostering a culture of cybersecurity awareness and providing ongoing training is essential for maximizing the effectiveness of new technologies (Pedreira et al., 2021).

14.6 RECOMMENDATIONS FOR FUTURE RESEARCH

As the landscape of cybersecurity continues to evolve, it is imperative that future research addresses the multifaceted challenges and opportunities presented by emerging technologies. This section outlines several key areas for future investigation, including longitudinal studies on cybersecurity effectiveness, comparative studies across sectors, and the exploration of hybrid cybersecurity solutions. One of the most pressing needs in cybersecurity research is the establishment of longitudinal studies that measure the long-term impacts of emerging cybersecurity technologies on KM systems. While many studies focus on short-term outcomes or specific incidents, understanding the sustained effectiveness of cybersecurity measures over time is crucial. Longitudinal research can provide insights into how organizations adapt to new threats, the evolving nature of cyber risks, and the effectiveness of various technologies in mitigating these risks. Such studies can also help identify best practices and inform the development of more resilient cybersecurity frameworks that align with the dynamic nature of cyber threats.

Another critical area for future research is the need for comparative studies that examine sector-specific cybersecurity needs and challenges. Different industries, such as healthcare, finance, and education, face unique cybersecurity threats and regulatory environments that influence their approach to data protection. For instance, the healthcare sector is particularly vulnerable to cyberattacks due to the sensitive nature of patient data and the increasing integration of technology in medical practices. Research that compares the cybersecurity strategies employed across various sectors can illuminate best practices, highlight common vulnerabilities, and provide tailored recommendations for organizations operating in diverse environments. This comparative approach can also facilitate knowledge sharing and collaboration among sectors, ultimately enhancing overall cybersecurity resilience. The exploration of hybrid cybersecurity solutions that combine multiple techniques is another promising avenue for future research. As organizations seek to bolster their defenses against increasingly sophisticated cyber threats, integrating approaches such as ZTA with artificial intelligence (AI) and blockchain technology may offer enhanced security capabilities. Research should focus on the efficacy of these hybrid models, examining how they can be effectively implemented and the potential challenges associated with their deployment. Additionally, studies could investigate the interplay between different cybersecurity technologies and how they can complement each other to create a more robust security posture.

The ongoing evolution of cybersecurity necessitates a proactive approach to research that addresses the complexities and challenges organizations face in safeguarding their KM systems. Longitudinal studies on cybersecurity effectiveness, comparative studies across sectors, and the exploration of hybrid cybersecurity solutions are critical areas that warrant further investigation. By advancing research in these domains, scholars and practitioners can contribute to the development of more effective and resilient cybersecurity strategies that are better equipped to address the dynamic nature of cyber threats.

14.7 CONCLUSION

This chapter has provided a comprehensive analysis of emerging trends and future directions in cybersecurity for KM systems. As organizations increasingly rely on digital platforms to store and manage their knowledge assets, the risks associated with cybersecurity threats continue to escalate. Through this analysis, we explored advanced technologies such as AI, machine learning, blockchain, and quantum computing, each offering unique solutions to fortify KM systems against cyber threats. Additionally, we discussed advanced cybersecurity frameworks, including ZTA, PETs, and automation strategies, all essential for safeguarding knowledge assets in today's complex digital landscape. The chapter also underscored the importance of ethical and regulatory considerations in implementing cybersecurity strategies, emphasizing the need for compliance with data privacy laws and the ethical use of AI-driven cybersecurity solutions. Addressing these aspects is critical for organizations aiming to build secure and trustworthy KM environments. However, the chapter identified several challenges, such as scalability, cost, and user adaptability, which must be addressed to fully harness the potential of these technologies. This chapter highlights the critical need for ongoing innovation, robust frameworks, and ethical responsibility in cybersecurity for KM. By adopting a proactive and integrated approach to these challenges, organizations can enhance their resilience, protect their knowledge assets, and sustain a competitive advantage. Future research should continue exploring hybrid cybersecurity solutions, sector-specific strategies, and longitudinal studies to adapt to the evolving threat landscape and to support effective and sustainable KM practices.

REFERENCES

Adegbite, A. O., Akinwolemiwa, D. I., Uwaoma, P. U., Kaggwa, S., Akindote, O. J., & Dawodu, S. O. (2023). Review of cybersecurity strategies in protecting national infrastructure: Perspectives from the USA. *Computer Science & IT Research Journal*, *4*(3), 200–219.

Althar, R. R., Samanta, D., Purushotham, S., Sengar, S. S., & Hewage, C. (2023). Design and development of artificial intelligence knowledge processing system for optimizing security of software system. *SN Computer Science*, *4*(4), 331.

Arafa, A., Sheerah, H. A., & Alsalamah, S. (2023). Emerging digital technologies in healthcare with a spotlight on cybersecurity: a narrative review. *Information*, *14*(12), 640.

Cao, Y., Pokhrel, S. R., Zhu, Y., Doss, R., & Li, G. (2024). Automation and orchestration of zero trust architecture: Potential solutions and challenges. *Machine Intelligence Research*, *21*(2), 294–317.

Coventry, L., & Branley, D. (2018). Cybersecurity in healthcare: A narrative review of trends, threats and ways forward. *Maturitas*, *113*, 48–52.

Determann, L. (2019). Healthy data protection. *Michigan Telecommunications and Technology Law Review*, *26*, 229.

Formosa, P., Wilson, M., & Richards, D. (2021). A principlist framework for cybersecurity ethics. *Computers & Security*, *109*, 102382.

Garba, A., Sirat, M. B., Hajar, S., & Dauda, I. B. (2020). Cyber security awareness among university students: A case study. *Science Proceedings Series*, *2*(1), 82–86.

Hosseini, S. M., Sikaroudi, M., Babaie, M., & Tizhoosh, H. R. (2023). Proportionally fair hospital collaborations in federated learning of histopathology images. *IEEE Transactions on Medical Imaging*, *42*(7), 1982–1995.

Komaruddin, K., Utama, A. S., Sudarmanto, E., & Sugiono, S. (2023). Islamic perspectives on cybersecurity and data privacy: Legal and ethical implications. *West Science Law and Human Rights*, *1*(04), 166–172.

Kour, R., Aljumaili, M., Karim, R., & Tretten, P. (2019). eMaintenance in railways: Issues and challenges in cybersecurity. *Proceedings of the Institution of Mechanical Engineers, Part F: Journal of Rail and Rapid Transit*, *233*(10), 1012–1022.

Malik, M. S. (2021). Role of artificial intelligence in cybersecurity improvement. *International Journal for Electronic Crime Investigation*, *5*(1), 43–51.

Morrow, P. J., & Fitzpatrick, T. M. (2020). US and international legal perspectives affecting cybersecurity corporate governance. *International Relations*, *8*(06), 231–239.

Norem, S., Rice, A. E., Erwin, S., Bridges, R. A., Oesch, S., & Weber, B. (2021). A mathematical framework for evaluation of SOAR tools with limited survey data. *European Symposium on Research in Computer Security*.

Pedreira, V., Barros, D., & Pinto, P. (2021). A review of attacks, vulnerabilities, and defenses in industry 4.0 with new challenges on data sovereignty ahead. *Sensors*, *21*(15), 5189.

Pokhrel, S. R., Li, G., Doss, R., & Nepal, S. (2023). Towards decentralized operationalization of zero trust architecture. *Authorea Preprints*.

Pooyandeh, M., Han, K.-J., & Sohn, I. (2022). Cybersecurity in the AI-based metaverse: A survey. *Applied Sciences*, *12*(24), 12993.

Pulyala, S. R., Desetty, A. G., & Jangampet, V. D. (2019). The impact of security orchestration, automation, and response (SOAR) on security operations center (SOC) efficiency: A comprehensive analysis. *Turkish Journal of Computer and Mathematics Education (TURCOMAT)*, *10*(3), 1545–1549.

Rangaraju, S. (2023). Ai sentry: Reinventing cybersecurity through intelligent threat detection. *EPH-International Journal of Science And Engineering*, *9*(3), 30–35.

Rizvi, M. (2023). Enhancing cybersecurity: The power of artificial intelligence in threat detection and prevention. *International Journal of Advanced Engineering Research and Science*, *10*(05).

Şandor, A., & Tonţ, G. (2021). Considerations regarding the inclusion of cybersecurity in knowledge management. *International conference knowledge-based organization*.

Sarker, I. H., Furhad, M. H., & Nowrozy, R. (2021). Ai-driven cybersecurity: an overview, security intelligence modeling and research directions. *SN Computer Science*, *2*(3), 173.

Stafford, T., Deitz, G., & Li, Y. (2018). The role of internal audit and user training in information security policy compliance. *Managerial Auditing Journal*, *33*(4), 410–424.

Taddeo, M., McCutcheon, T., & Floridi, L. (2019). Trusting artificial intelligence in cybersecurity is a double-edged sword. *Nature Machine Intelligence*, *1*(12), 557–560.

Walkowski, M., Krakowiak, M., Oko, J., & Sujecki, S. (2020). Efficient algorithm for providing live vulnerability assessment in corporate network environment. *Applied Sciences*, *10*(21), 7926.

Wang, J., Wang, Y., Song, J., & Cheng, H. (2023). IoV vulnerability classification algorithm based on knowledge graph. *Electronics*, *12*(23), 4749.

Wang, S., & Wang, H. (2021). A sociotechnical systems analysis of knowledge management for cybersecurity. *International Journal of Sociotechnology and Knowledge Development (IJSKD)*, *13*(3), 77–94.

Zhang, T., He, C., Ma, T., Gao, L., Ma, M., & Avestimehr, S. (2021). Federated learning for internet of things. *Proceedings of the 19th ACM Conference on Embedded Networked Sensor Systems*.

Zimmerman, T. A., & Zimmerman, T. A. (2019). *Manufacturing profile implementation methodology for a robotic workcell*. US Department of Commerce, National Institute of Standards and Technology.

Kolibri 360 GmbH
Ferdinand-Jühlke-Straße 7
99095 Erfurt, Deutschland
produktsicherheit@kolibri360.de